PRAISE FO

'A rousing and valuable[...]
should read' – F[...]

'*Too Much Luck* is a timely book, one that taps into important
public debate … Cleary also brings to the task a sense of
conviction and compelling advocacy.' – ROBERT MURRAY,
The Weekend Australian

'This is a carefully researched and well-written warning.
The statistics are frightening.' – BRUCE ELDER,
The Sydney Morning Herald

'A timely and provocative analysis of some of the risks
and opportunities associated with the present resources boom'
– SAUL ESLAKE, *The Monthly*

'Well-written and thoughtful' – ANNABEL CRABB, *The Drum*

'A call for action … it presents a powerful and passionate
case for more sophisticated management of our mineral wealth …
In doing so, [Cleary] does great service to all Australians.'
– MICHAEL GILDING, *The Canberra Times*

MINE-FIELD

MINE-FIELD

THE DARK SIDE OF AUSTRALIA'S RESOURCES RUSH

PAUL CLEARY

Black Inc.

RICH
AU

Published by Black Inc.,
an imprint of Schwartz Media Pty Ltd
37–39 Langridge Street
Collingwood VIC 3066 Australia
email: enquiries@blackincbooks.com
http://www.blackincbooks.com

The National Library of Australia Cataloguing-in-Publication entry:

Cleary, Paul, 1964-

Mine-field : the dark side of Australia's resources rush / Paul Cleary.

ISBN: 9781863955706 (pbk.)

Includes bibliographical references.

Mines and mineral resources--Environmental aspects--Australia.
Mines and mineral resources--Social aspects--Australia.
Mines and mineral resources--Economic aspects--Australia.

333.7650994

Printed in Australia by Griffin Press. The paper this book is printed on is
certified against the Forest Stewardship Council® Standards. Griffin Press
holds FSC chain of custody certification SGS-COC-005088. FSC promotes
environmentally responsible, socially beneficial and economically viable
management of the world's forests.

FSC
www.fsc.org
MIX
Paper from
responsible sources
FSC® C009448

CONTENTS

This book is dedicated to the high-school teachers I was lucky to have: Paul Oakley cfc, Cliff Fogarty cfc, Harry Bahr, Paul Doney, Paul Ryan and especially my economics teacher, the late Paul Cheney.

PREFACE

For more than 150 years, the resources industry, both mining and energy, has coexisted with the rest of the Australian economy and society. For the most part it has been a positive influence, especially in regional and remote areas, where mining has been a catalyst for development despite sometimes leaving large dollops of degradation behind. But that was when mining and energy were minor parts of our economy, and when our democratic institutions and good financial management helped us to avoid some of the nasty side-effects of resource rushes experienced overseas, most notably in the developing world. Now things have changed.

In less than a decade, the frenzied pace of Australian resource development has tipped the balance of coexistence to the point where mining dominates our society, our economy and even our political system. The revenue earned by mining companies has already tripled as a share of our economy, from around 5 per cent for most of the post-war era to 15 per cent.[1] Even on this scale, the industry has been able to dictate terms to compliant state and federal governments. It has erased small communities and towns; occupied vast tracts of prime farmland; constructed ports and liquefied natural gas (LNG) plants in the Great Barrier Reef Marine Park; diverted entire rivers; and built on top of 30,000-year-old sacred sites. It has even helped remove an elected prime minister from office. If this is what the industry can do now, how much more power will it wield if it rises to one-fifth or even one-quarter of national income?

Such a rise depends on the direction of commodity prices, but certainly the industry is gearing up to expand massively the volume of production. Having already doubled iron ore production since the beginning of the boom, global and local companies are preparing to double output again over the course of this decade, possibly reaching a mammoth 1 billion tonnes a year by 2020.[2] This iron ore production would be equivalent to all of Australia's mineral output in 2011. Likewise, coal production has increased by about 60 per cent since 2002, but output could double this decade. Meanwhile, gas companies have already doubled LNG export volumes and are on track to quadruple them by 2020. Soaring prices and profits have made the resources companies hell-bent on expansion, transforming Australia from a country with a diversified economic base and relatively healthy democratic institutions into one dominated by the interests of global mining giants. In the future, Australia will be even more dependent on volatile resource exports for its foreign income. These have already doubled in the space of a decade to account for two-thirds of our export income. As the high dollar squeezes out other exporters, the share could go considerably higher.

The human cost of this expansion seems too high a price to pay. The mining juggernaut has rolled over communities in the Hunter Valley and the Darling Downs, and many more will be wiped off the map. Those not lucky enough to be bought out are left to endure the assault – dust, noise, blasting, gases and lights as bright as at a football stadium – of the 24/7 operations occurring at most mining and coal-seam gas (CSG) operations. The expansion is driven by gruelling rosters that involve 12 hour–plus shifts and fortnightly spells away from family and community. Rising road carnage is another result of the long drives that workers are forced to take at the end of their shifts.

The resources industry has an insatiable demand for new territory and is encroaching onto our best farmland, imposing on farmers and rural communities like never before. The cost of giving these companies such sweeping access to our resources could be permanent damage to Australia's food bowl and water resources – a high price to pay for the driest continent on earth. Yet our state governments, which have the biggest say over resource developments, are remarkably enthusiastic about handing out mining concessions in these areas, largely because they get an upfront cut of the earnings.

Little stands between the people and the resources rush. Long-time farmers who have been inundated with government inspectors throughout their careers, and who now have mines or drills on their land, say they've never had an official come to check that CSG wells have been properly drilled, or to measure the dust, noise and fumes stirred up. Governments do surprisingly little to monitor the harmful effects of mining, leaving the task almost entirely to the companies involved. Much of this reporting seems inadequate or biased and is kept confidential; even when companies exceed safety limits, repercussions are rare. Our governments are ill-prepared and, it seems, unwilling to manage and monitor future expansion.

If Australia is to avoid the pitfalls of resource-sector dominance, then serious regulatory reform is needed. Only then will we have first world governance over the extraction of the resources that belong to all Australian citizens – and to future generations. Our political leaders need to think and act as the masters rather than the servants of the companies making a handsome profit from our common wealth. In other words, our leaders need to think and act more like Arab oil sheiks than hapless clients.

Reform is desperately needed in three key areas: project regulation, taxation and minimising the human impact. Firstly, most project regulation sits in the hands of weak state governments that are prejudiced in favour of development. Companies generate their own Environmental Impact Statements (EISs) without any independent review or quality control. The capacity of states to oversee developments is being undermined by companies that poach their best officials. Australia needs a well-resourced federal 'super regulator' to be run jointly with the states. This regulator would manage the environmental assessment process and deliver real-time monitoring, instead of filing away confidential reports.

Second, there is an imbalance at the heart of Australia's economy and society: we are a first world country with third world rates of taxation on exploitation of our national resources, a situation that has brought about a tsunami of investment to exploit them. The federal government's mineral resource rent tax won't make much difference; in fact, from the time the tax was announced in July 2010, to April 2012, just after it became law, the value of major resource projects approved by company boards for development had more than doubled to $261 billion. State governments have shown they have no idea how to tax the resources sector, as they insist on retaining an archaic system of royalty payments that fails to capture a share of profits. Even worse, these production-based royalties penalise small domestic companies at the expense of big multinationals and have become a form of inducement that leads to distorted decision-making.

Finally, companies should be made responsible for their impact on people living on the edge of resource projects, and they should also be forced to minimise the impact of work regimes that border on the inhumane. Fly-in fly-out (FIFO) rosters are destroying marriages and communities, and companies are in denial about the road accidents caused by exhausted workers.

It is not a case of governments and companies putting royalties and profits before people; instead it is as though people don't matter at all. Presently, regulation is focused more on flora and fauna than on the people affected by mining and energy developments.[3] Some of these people live on the edge of ever-expanding mines or close beside the railway lines that ship ore from mines to ports, or near exposed coal dumps and loading facilities. Many live with the knowledge that a huge mine or CSG project is about to open up on their land and they have no negotiating power, let alone a right of veto, in the face of the government's enthusiastic approval. Others are living with the memory of their land, perhaps held by their families for generations, after a mining or energy company has given them no choice but to sell. Indeed, so great, so far-reaching and so uncontrolled is the resources boom that it appears to be driving the biggest forcible transfer of land on this continent since the first wave of white settlement.

In my previous book, *Too Much Luck*, I looked at the big picture of the resources boom – the rush to invest and the failure to tax, and the need for a sovereign wealth fund to save windfall revenue for when the boom ends, and for when the resources start to run out. In *Mine-Field* I get down in the dirt to count the human and economic costs of Australia's mineral addiction, while also looking for solutions to better manage the once-only development of our resources for the benefit of all parts of society and our economy, and for the benefit of future generations of Australians.

INTRODUCTION

SUPER-SIZED MINING

On a sun-drenched spring afternoon, the undulating fields of the black soil country just west of the Great Dividing Range lie dotted with rolls of hay after yielding a bumper crop, while freshly ploughed paddocks resemble an expanse of chocolate cake as they await the summer sowing. This oasis of rich black soil on the world's most arid continent oozes fertility, from intensive market gardens on the top of the range to expansive grainfields and grazing paddocks on the Downs to the west, producing a cornucopia of vegetables, grains and beef. To the local farmers it is known affectionately as 'sweet country'.

This food bowl extends across Queensland's Darling Downs, west of the 'garden city' of Toowoomba, and the productivity of this foodbowl is matched, if not exceeded, by the Liverpool Plains region in northern New South Wales. It feeds the nation and millions more around the world. The sustainable practices embraced by farmers could make this country productive for generations to come – but the farms are by now being overrun with breathtakingly huge mining and gas projects. Almost overnight, a farmer's paradise is being turned into an industrial landscape.

On the edge of Toowoomba an industrial zone to service the gas industry is being carved out of the black soil paddocks, and 12 kilometres further west the proposed Charlton Wellcamp industrial estate is wiping out farmland for a 2,000-hectare estate destined to become a heavy industrial zone for the resources industry. The local council plans to seek a special exemption from state pollution laws for the industrial estate. Planning maps for the region show a spaghetti pattern of proposed railway lines, high-voltage power lines, petroleum pipelines and heavy haulage roads reaching out to future developments, some yet to be revealed. Another hour's drive to the west, near the town of Dalby, invasive networks of gas pipelines, access roads, pressurisation stations and waste water storage ponds are spreading out over a vast area. Mining and CSG wells are changing this rural landscape at lightning speed.

Amid this industrial onslaught is the 3,200-hectare spread operated by Katie and Scott Lloyd near Tara, 300 kilometres west of Brisbane, one of the many gems of Australian agriculture. A sophisticated mixed farming operation, it runs about 5,000 cattle while rotating summer and winter crops on 1,250 hectares of laser beam–leveled paddocks. Take one look at this energetic mid-thirties couple and their two young boys and what you see is a reassuring image for the future.

As farmers engaged in food production, the Lloyds accept that they have to comply with a raft of regulations. They don't seem to mind the surprise visits from government inspectors who check that their feedlot is being properly maintained, or that their weighing equipment is accurate. They have even come to accept the restrictions on their allocation of bore water drawn from an aquifer that is part of the vast underground resource known as the Great Artesian Basin. Nor do the controls end when the cattle go out the gate – ear tags ensure that the meat can be traced straight back to the farm.

Yet all of this is in stark contrast to the oversight applied to the risky CSG projects being rolled out at breakneck speed on prime farmland in Queensland and New South Wales, including on the Lloyds' farm, and the regulations imposed on mining in general. In the past three years, the Lloyds have had 18 CSG wells constructed on their property, which came about largely because the Lloyds had little choice but to sign. Under state law, gas companies can get access to this prime land after a minimum negotiation period of just 20 business days. This allows them to carry out 'advanced activities' like drilling, and after just 50 days they can obtain an Entry Notice from the Land Court, which provides full access rights to develop their production wells.[4]

In the three years that the 18 wells on the Lloyds' property have been in production, no-one from the government has undertaken random inspections to check that the APLNG joint venture between Australia's Origin Energy and US giant ConocoPhillips is doing the right thing by the Lloyds or by the wider community that relies on the aquifers that sit above the coal seams. Nor did anyone from the government come onto the Lloyds' land to examine what happened with the two exploration wells drilled by Queensland Gas Corporation (QGC) when they malfunctioned.

'You could see gas coming up through the surface of the ground. When water pooled, it bubbled,' says Katie Lloyd.[5] In August 2010, government inspectors from the Department of Environment and Resource Management had wanted to inspect the wells on the Lloyds' property but, according to Lloyd, QGC told the government 'they couldn't access the property because there had been issues with us [the Lloyds] as landowners'.

'We were very concerned by this and it certainly had us asking questions as to why CSG companies could dictate the terms of supposed "random inspections" to the regulators,' Lloyd says. QGC confirmed that two of its wells produced 'minor seeps' of low-pressure gas. 'The seeps posed no unacceptable safety risk,' the company said. QGC said it had 'no knowledge' of having advised the state government not to visit the property.[6]

The Lloyds were so concerned about the lack of proper governance over the CSG operations on their land that they raised the issue in a meeting with Queensland's then premier, Anna Bligh, in November 2010, but the inspectors didn't come until early 2011, after the two malfunctioning wells had been removed.

This lack of government oversight is just one of the many worries for the Lloyds and other farmers living with the CSG and mining projects mushrooming all over Australia. In the short time that CSG companies have been operating in their region, the water levels in the Lloyds' bores have been steadily dropping. While water restrictions are imposed on the Lloyds and their neighbours, CSG companies will be able to draw limitless amounts of water from the coal seams beneath farmland, even though this water is saline and the companies are yet to develop a workable plan for dealing with it. After being told by the government that their farm can have no more water, Katie Lloyd says she finds it 'unbelievable' that the CSG company operating on the land can take as much water as it likes, 'purely because it falls under the Petroleum and Gas Act'.

Then there's the constant stream of company employees who, without asking permission, come onto the Lloyds' land to do maintenance work, contrary to what APLNG said during negotiations. According to the Lloyds, APLNG told them that the visits 'would eventually be weekly, if that much at all'.

The gas boom has changed life demonstrably for the worse for the Lloyds and many families who have tilled the black soil of the Darling Downs for generations. Soaring demand for labour for the building of these mammoth projects has driven up the Lloyds' wages bill for farm employees more than 50 per cent in the space of a few years, more than offsetting the income that farmers get for hosting the CSG wells. And the compensation the Lloyds receive is based only on the amount of land used by the project, not the profits or income earned by disrupting their lives and their farming operation. There's land on their farm that can't be accessed for cultivation because of the 38 water- and gas-release valves, known as 'high point vents' and 'low point drains', above the buried pipeline. These valves are encased in concrete shells situated along an access road, preventing machinery like harvesters and wide-load farming equipment from travelling to the nearby fields.

Most of the properties around the Lloyds' have been bought out by CSG companies, bringing thousands of men to live in work camps in a 10-kilometre radius of their farm. Once family-friendly pubs are now crowded with poker machines and men in fluoro shirts.

And yet the resources boom that's already transformed southeast Queensland has only just begun. Companies have installed less than a tenth of the projected 40,000 production wells expected to be rolled out over 20,000 square kilometres of prime farmland. The Lloyds thought they'd done their bit for the boom, but now they face the prospect of another 30 wells. Unlike their current 18 wells, some of these would be placed in the immediate vicinity of their house, feedmill, feedlot and sheds, directly affecting both their business and their lifestyle. The Lloyds fear these new wells will severely constrain their ability to expand and diversify their business in the years ahead, yet the wells will go in without their permission should QGC decide to proceed. The Lloyds have good reason to be worried. This company, a subsidiary of the UK giant BG Group, has employed tough tactics with farmers in the area, such as insisting on putting major pipelines underneath cultivation paddocks and thereby limiting their use, rather than going around them. Then there's the prospect of hydraulic fracturing, or fracking – the use of chemicals, sand and water blasted under extreme pressure to release the gas trapped in the coal seams – which could contaminate water from the aquifers that the Lloyds and their children drink, along with their 5,000 cattle. However, QGC said it had no current plans to develop wells on the Lloyds' property. It said there were 'many instances of coexistence between farming and the gas industry where infrastructure such as pipelines crosses cultivation areas and cultivation continues'. [7]

The Lloyds' experience is typical of people who live near resource projects. While governments boast of rigorous environmental assessments and numerous conditions when they approve these projects, by and large companies are allowed to self-regulate. Companies commission their own monitoring of dust, noise, heavy metals, and toxins like arsenic, sulphur dioxide and nitrous dioxide; people living near mines say these readings are often taken when conditions like wind and rain are most favourable. Far too often, companies blame factors like equipment failure when it suits them. Even when they use independent consultants, there is a bias towards generating favourable results. Companies are able to average out their readings over long periods, thereby eliminating times when they have exceeded their limits. And when occasionally they give reports to the government showing that they have exceeded their limits, most of the time the reports get filed away and nothing happens.

Such lapses in regulation would have been concerning when mining and energy played a relatively minor role in our economy, but they become alarming given the enormous scale of projects being rolled out. The size and speed of development increases the risks to natural resources and

Australians, and yet the words 'asleep at the wheel' are too mild to capture the lack of responsibility taken by governments for protecting the public interest.

The Lloyds are among tens of thousands of people in rural Australia living in the shadow of Australia's resources rush. They are decent and hardworking people who produce the food that most of us take for granted. But in this latest resources boom, the fifth in our history, governments hooked on the inducements – known as royalties – paid by resources companies seem willing to treat families like the Lloyds as expendable.

MAGIC PUDDING

The frenzied development creating upheaval for communities all over regional Australia is part of the so-called resources 'super cycle', a wave of investment tied to China and India's industrialisation. Australia's top economic advisers in Treasury and the Reserve Bank are confident that this boom, unlike those before it, will transcend the business cycle and that it could span several decades. That might seem a heroic forecast, but so far the boom certainly has been bigger and longer-lasting than any of those last century. Despite the ongoing global turmoil, Australia remains a highly profitable place from which to extract minerals and energy, which is why investment continues to grow amid dark economic stormclouds.

The super cycle is in the process of delivering an unprecedented number of very big, or super-sized, mining projects.

In mid-2012, Australia had 373 active mining complexes, but the size of most is being expanded and the overall number is growing swiftly. As noted, plans are in place to more than double coal and iron ore production, and quadruple LNG production, by the end of this decade. The federal government's official list of major resource projects, defined as those with capital expenditure of more than $40 million, and approved by company boards and governments for construction, currently has 98 such projects, involving record capital expenditure of $261 billion.[8] By comparison, at the height of the boom last decade there was $70 billion worth of projects. Ministers call the list 'the investment pipeline'; with Freudian overtones, they claim to have the biggest such pipeline in the world.

And there's more. The government's investment survey identifies another 300 less-advanced projects worth more than $240 billion that are either undergoing feasibility studies or awaiting the outcome of government approval processes. Even though these projects are only at the planning stage, federal ministers are so excited by the prospect that they talk about Australia having a $500 billion pipeline of resources investment. This approved and projected investment exceeds the total amount spent by companies since the start of this boom, indicating that it has only just begun.

The Australian investment pipeline reaches into just about every corner of the country, but especially into farming regions on the east coast, like the coal country of the Hunter Valley, north of Sydney, the black soil plains around Narrabri and Gunnedah further north, and then onto the

Darling Downs just over the Queensland border, and the grazing country of the Surat and Galilee basins. These massive mining and energy projects are being rolled out with military precision and concentrated in three states. With 41 such projects, Western Australia is regarded as the powerhouse resource state, but Queensland and New South Wales are not far behind, with 29 and 18 projects respectively. However, there's a stark difference between the east and west coast booms. In Western Australia, most of the projects are being built in remote areas like the Pilbara; in Queensland and New South Wales, this investment intersects with prime farmland.

The big new variable in this mining boom is not the number of projects, but their scale. The trebling of commodity prices since 2004 has allowed resources companies to garner much more investment capital, and their plans have thus become ever more ambitious. One emerging trend is the planning of breathtakingly large mines; another is the transformation of mines that were approved as modest into mega open-cut mines. It is no exaggeration to say that just about every one of the 373 operating mines in Australia is being expanded, some of them dramatically so. Increasingly, what used to be regarded as an exceptionally large mine is becoming commonplace. It's as though the companies have discovered a magic pudding. Higher prices allow mines to be enlarged with feats of engineering that push the boundaries of technology and prudent management. These plans will hasten the depletion of Australia's known resources, but the race for riches overshadows this fact.

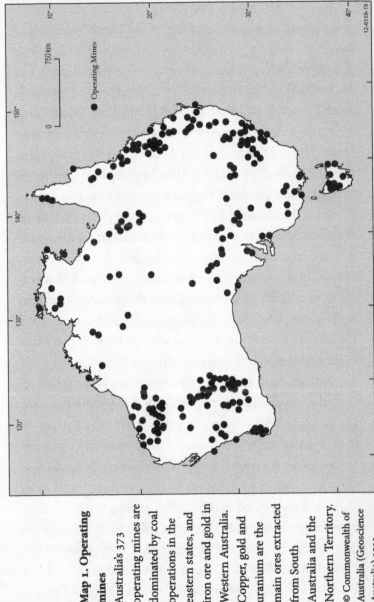

Map 1. Operating mines

Australia's 373 operating mines are dominated by coal operations in the eastern states, and iron ore and gold in Western Australia. Copper, gold and uranium are the main ores extracted from South Australia and the Northern Territory. © Commonwealth of Australia (Geoscience Australia) 2012.

750 km

• Operating Mines

12-6139-19

A decade ago, a mine that produced 10 million tonnes of coal per year was considered one of the largest such operations in the country. Ten million tonnes is enough to fill 33,000 of those yellow 300-tonne trucks and stretch them in a line for 500 kilometres. The Hunter Valley mining town of Muswellbrook has one such operation, BHP Billiton's Mt Arthur Coal mine, one of several ever-expanding pits that abuts the fast-growing town of 10,000 people. Located just five kilometres south of Muswellbrook, Mt Arthur has long held the title of the biggest coal mine in New South Wales. It has existed since the 1960s, but the three-fold increase in coal prices since 2004 prompted BHP to expand output to 15 million tonnes a year in 2009, earning it the title of the biggest coal mine in the country, and now another expansion plan will push that output to 24 million tonnes a year. Muswellbrook's deputy mayor, Jeff Drayton, who works at the mine, proudly boasts of Mt Arthur's status, but the title may not last long, as there is a queue of new mines of gargantuan proportions.[9]

Western Australia's iron ore mines are ramping up to more than double production while in Queensland there are advanced plans for mines destined to produce more than 30, 40 or even 60 million tonnes of coal a year. Queensland's former Labor government and its new Liberal–National coalition have shown a willingness to approve these mines even when they are located in productive farming regions. Immediately after the landslide March 2012 election in Queensland, the Queensland Land and Environment Court rejected a bid to block Xstrata's Wandoan coal mine, which is located 350 kilometres northwest of Brisbane. The mine has an estimated

1.2 billion tonnes of coal and, at a production rate of 30 million tonnes a year, could be in operation for 40 years. It will consume 300 square kilometres of farmland, including grazing and cropping land.

Upon winning office in March 2012, Premier Campbell Newman ruled out two mining projects that were proposed by small Australian companies, but he said he supported the Wandoan proposal approved by the former Labor government. 'Wandoan is a different matter,' said Newman, after the land court's decision, adding that unlike the two mines he opposed, Wandoan wasn't included in the state's definition of strategic cropping land, even though it produces grain and beef.

Perhaps Newman just couldn't say no to Swiss-based Xstrata, which along with BHP Billiton and Rio Tinto has in recent years displayed devastating power and influence by helping to knock off the super-profits tax and Prime Minister Kevin Rudd. In April 2012, Xstrata was taken over by its major shareholder, Glencore, a commodities trading house founded by Marc Rich, who fled to Switzerland after being charged by the US government for tax evasion. Glencore's modus operandi is to profit from control of the supply of key commodities, and it has been accused of creating speculative bubbles.

One Wandoan livestock and grain farmer, John Erbacher, says his 1,000-hectare property was among the best in Queensland, yet he faces having an explosives store on his farm with a 600-metre buffer zone around it. There are many more Wandoans on the way, and there will be hundreds more farmers like Erbacher.

In a virgin coal precinct known as the Galilee Basin, located about 400 kilometres directly inland from Gladstone, Queensland, Australia's iron lady, Gina Rinehart, has conditional approval for the first of two giant mines aimed at extracting 8 billion tonnes of coal. In partnership with India's GVK, the 24 kilometre–long open cut pits of the Alpha mine will yield 30 million tonnes a year, and the joint venture plans a second mine of the same capacity. Clive Palmer has a deal with Chinese state-owned enterprises to develop a 40 million tonne-a-year longwall mine that threatens the 8,000-hectare Bimblebox Nature Refuge. But India's Adani Group takes the prize for the single biggest mine, with plans for a 60 million tonne-a-year facility that the company says could be in operation for a century. In yellow truck terms, Adani's operation would fill 200,000 every year; and if lined up bumper to bumper, they would stretch for 3,000 kilometres, from the mine at Alpha all the way to Collins St, Melbourne, and back again. All of this coal will be shipped along a new 500 kilometre–long railway line and then piled into bulk carriers at an expanded Abbot Point terminal, before sailing through the Great Barrier Reef, one of the greatest marine wonders in the world. With these exports generating thousands more ship movements a year, there's likely to be many more bulk carriers like the *Shen Neng 1*, which ploughed into the reef off Great Keppel Island in April 2010, inflicting a three kilometre–long scar on the reef before spilling four tonnes of oil.

Unlike previous booms, where the mines helped to develop regional and remote areas by building towns and cities, these new projects have an eerie sameness about them.

They increasingly involve FIFO workers who live in army-style barracks made up of aluminium sheds known as dongas for two weeks or more at a time, before flying home. The mining company magnates, who look as though they enjoy super-sized burgers, are adopting the McDonald's approach to development.

The impact of these projects is enormous and hard to predict because they involve cumulative effects: the combined impact of a cluster of mines is greater than that shown in the reports produced for each individual mine. This will be especially the case in places like the Galilee Basin, where the Palmer, Rinehart and Adani mega-mines are planned.

The capacity of state and federal governments to regulate this development is surprisingly weak. Unlike best practice overseas, most regulation isn't handled by independent statutory authorities that dispassionately and professionally assess the merits and risks of proposed developments. Such approvals are handled by state government departments that report directly to ministers, whose governments have a financial interest in approving new developments. Because state governments are hooked on the royalties they collect from mines, they rarely say no.

FOOTPRINTS OF GIANTS

When a mine goes ahead, it generates a ripple effect that spreads far beyond the pit. To start with, there's the mine itself, the land it consumes, the aquifers that are dug up, and the surrounding buffer zone. Mine development often means sacrificing prime farmland, forests abundant with biodiversity, and land rich in Aboriginal heritage. Then there's the impact on the people located in the immediate vicinity, who have to live with the dust, noise, bright lights and plumes of toxic gas.

When there are no railway lines nearby, the roads radiating from the mine carry enormous double or triple semi-trailers used to ship the ore, and a steady stream of fuel tankers to supply diesel to the energy-hungry operations. Roads are also made busier and more dangerous by the shift-workers who often drive long distances in and out of the mine. Road accidents in Queensland's Bowen Basin region, especially those involving heavy vehicles, are up 60 per cent since the start of the mining boom mid last decade.

When the ore is shipped by rail, it often gets dumped on sidings before being loaded onto wagons; it then passes through urban areas on the way to the port, usually spreading dangerous coal dust because the wagons are unsealed. Although governments attempt to limit the environmental damage of mines by requiring companies to produce Environmental Impact Statements (EIS), there's very little analysis of the human impact. Multiply this by the hundreds of mining and energy projects being rolled out, and the result is a rolling transformation of cities, towns and farming regions.

The billionaire mining magnate Clive Palmer makes the audacious claim that mines cannot be compared to the expansive CSG projects. 'Unlike coal seam gas, a mine is not something that's going to destroy hundreds of acres of land … I don't think we have anything to worry about coal mining itself, because a mine only takes over a defined area where the pit will be.'[10] Yet that defined area is very, very big indeed.

The EIS for Palmer's China First Coal Project states that the mining lease makes 21,561 square kilometres available for coal exploration, including the Bimblebox reserve, which is protected from every sort of encroachment except mining. This area covered by the China First lease is in fact larger than Kakadu National Park, Australia's biggest national park. The EIS states that 85,000 hectares of 'good quality' agricultural land will be lost 'for the duration of mine operations', with no guarantee that such land will be restored, while a further 100,000 hectares of lower-quality land will be affected. In addition, the China First EIS states that 'it is anticipated that the regional availability of Class C agricultural land will be further impacted by this project'.[11] But this is just the beginning. The EIS declares breathlessly that the Galilee Basin is so rich in coal that it could result in an 'ultimate scenario' of 400 million tonnes a year, as much as Australia's total annual coal production in 2012, and all of this would have to be shipped through the Great Barrier Reef.[12]

This is not minimal-impact mining, and these are not mere mines. They are mega-mines, and when three or four of them are put together, then it is mega-mining to the nth degree. Environmental scientists refer to the unpredictable nature of cumulative effects of multiple mines as they can accumulate linearly or exponentially and reach 'tipping points' after which major change in ecosystems may follow.[13] As a result, the impact of multiple mines is becoming more difficult to predict. China First's EIS lists no fewer than 71 proposed mines in its immediate vicinity.[14] The cumulative impact of these projects would affect groundwater especially, because 'open cut mining involves the removal of significant volumes of overburden and target material resulting in significant open voids,' the EIS states. It adds: 'The open voids are likely to significantly alter the hydro-geological regime of the aquifers they intersect as they act as artificial sinks for groundwater.'

The loss of water resources is a significant issue for our arid nation. Numerous farmers in the Hunter Valley, who have experienced the rollout of mines since the 1980s, report the loss of springs and aquifers as a result. And all over Western Australia's Pilbara region, Aboriginal people say that many ancient springs that have flowed for tens of thousands of years have recently gone dry.[15]

*

In a moment of candour, perhaps aiming to deflect attention from his own mega-mine, Palmer claimed that CSG uses 'lethal' technology that 'will kill Australians, poison our water table and destroy the land'.[16]

Vast areas of rich farmland and crucial water resources are affected by CSG, as the Lloyd family can testify. The footprint is of a scale never before seen in our country's history. But so too is the footprint of the mega-mines planned for farming regions in Queensland and New South Wales, and the grazing country of South and Western Australia. If we look at the coal and CSG rushes together, it is clear that they are destined to change the face of regional Australia forever.

PART 1: TWO RUSHES

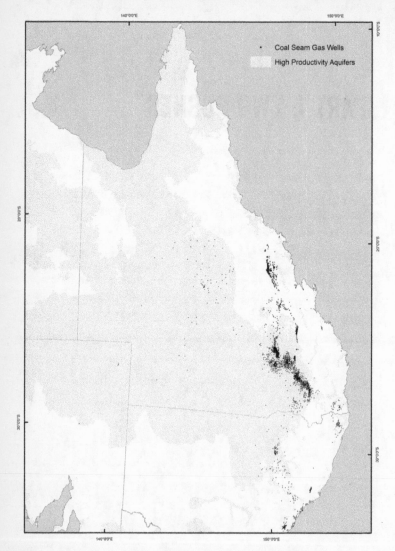

Map 2. Coal-seam gas wells and high-productivity aquifers

5,000 of the projected 40,000 wells drilled into some of the best farmland in Queensland and New South Wales.

Source: Geoscience Australia Datasets – Mine and resource locations and aquifer data.

SPIDERWEBS AND ANT NESTS

When seen for the first time from the air, it looks as though a giant spiderweb has spread over what was until recently a vast patchwork of paddocks and fields. Like nothing ever before visited upon the continent of Australia, a network of wells, access roads, pipelines and pressurisation stations is fanning out across the Darling Downs. This is what coal-seam gas looks like from above; beneath the surface it is even more intrusive, akin to a vast network of ants' nests fanning out deep below the earth's surface, all connected to their ant Queen, the pressurisation station.

What can now be seen in Queensland is the result of the 5,000 CSG wells installed so far, but this is just one-fifth to one-tenth of the likely number to be drilled into agricultural land in the state by mid-century. Only 2,400 of these 5,000 wells are actually producing gas; the balance are either appraisal or exploration wells. In what is arguably the biggest engineering feat since the Snowy Mountains Scheme, three mammoth projects worth $70 billion are being built over 20,000 square kilometres of good cultivation and grazing land.

These projects alone involve drilling almost 20,000 production wells, but another five in development could push that number to 40,000 over coming decades.[17] There remain many

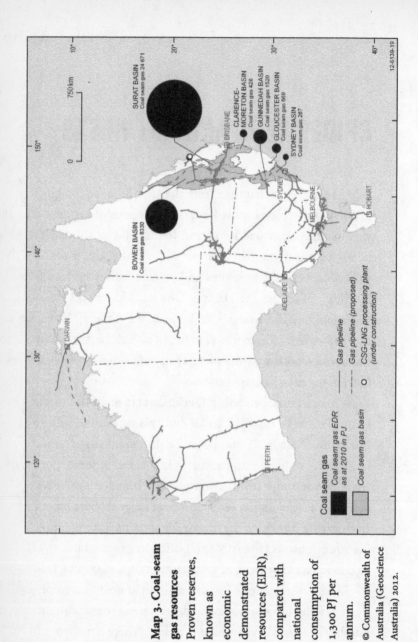

Map 3. Coal-seam gas resources

Proven reserves, known as economic demonstrated resources (EDR), as at 2010, compared with national consumption of 1,300 PJ per annum.

© Commonwealth of Australia (Geoscience Australia) 2012.

Coal seam gas

■ Coal seam gas EDR as at 2010 in PJ

▨ Coal seam gas basin

— Gas pipeline

--- Gas pipeline (proposed)

○ CSG-LNG processing plant (under construction)

SURAT BASIN
Coal seam gas 24 671

BOWEN BASIN
Coal seam gas 8330

CLARENCE-MORETON BASIN
Coal seam gas 428

GUNNEDAH BASIN
Coal seam gas 1520

GLOUCESTER BASIN
Coal seam gas 569

SYDNEY BASIN
Coal seam gas 297

DARWIN

PERTH

ADELAIDE

BRISBANE

SYDNEY

MELBOURNE

HOBART

750 km

12-6139-19

unknowns about how these developments will interact with the groundwater systems that make up the Great Artesian Basin. Given the uncertainties, other states have been more cautious. New South Wales has put a moratorium on the technique of fracking and on the building of evaporation ponds for waste water, and so far has granted approval for only 275 wells to be installed, but this has not halted ambitious exploration drilling, including some by Dart Energy just five kilometres from the centre of Sydney.[18]

Unlike in the 1960s and 1970s, when gas companies developed fields like the Cooper Basin in remote parts of the country, far away from populations centres, groundwater and prime farmland, the Queensland and NSW governments have now granted licences for CSG extraction on some of the best farmland in Australia, and close to or even within towns and cities. In the rush to rake in the royalties, these governments have carpeted their states with exploration licences, while Queensland has expedited full-scale production.

CSG developments pose a range of serious threats to groundwater, farmland, farming families and rural communities, and to the national economy. Based on overseas experience and the risks identified by developers and regulators in Australia, the industry has the potential to do the following: contaminate underground aquifers; produce billions of litres of unmanageable saline waste water that will yield millions of tonnes of salt and threaten farmland, river systems and wetlands; overlay an extensive network of access roads and pipelines; accelerate climate change by leaking methane gas into the atmosphere; trigger earthquakes; depress land values; and impose a crippling cost across the economy by doubling or even tripling the price of domestic gas.

Like all resource projects, CSG developments have inherent risks, but the likelihood of at least some of these dangers being realised, and indeed multiplied, is heightened by their enormous size and what seems to be the diminishing capacity, or perhaps willingness, of state governments to regulate them effectively.

When the Queensland projects were on the horizon in 2006, Geoff Edwards, a principal policy officer with the state Department of Mines and Energy, raised the alarm about the capacity of industry and government to manage the potential downsides. Edwards' paper was confined to the topic of managing waste water produced by CSG – just one of the challenges posed by this form of energy extraction. He warned there was an attitude of 'technological optimism' about how to deal with the problem, and that the industry had been 'developed faster than the capabilities of the authorities to moderate the potential downsides'.[19] Edwards also warned that the simultaneous approval of multiple, large-scale projects could make matters even worse – which is exactly what occurred. Edwards spelt out the cumulative impact in blunt terms:

> Each company may well be able to manage its own patch, but each company's efforts will not be adequate to manage the interplay of effects once many new fields are added and flood vulnerability, roadworks, possible new dams, changes in the grain industry and climate change considerations are superimposed.[20]

Five years later, some of the key players involved in sanctioning these developments, including the environment minister, Tony Burke, and the resources minister, Martin Ferguson, echoed Edwards' warning. In early 2011, Ferguson had declared that the CSG developments in Queensland represented a 'new industry for Australia' that positioned the country to become the second-biggest LNG exporter in the world. By the end of that same year, however, he conceded that 'the industry has grown too quickly', although he believed it should 'concentrate a little more on community engagement' to overcome this problem.

While the potential risks from CSG developments cannot be quantified, companies are serious about ensuring that one of them – a sharp increase of domestic gas prices – is indeed realised. Gas companies have been signalling to each other about their imminent plans to ramp these up as soon as they start exporting. This is because domestic gas prices are about half to one-third the price of the Asian LNG market, so linking the two markets will allow the companies to apply world parity pricing.

In a November 2011 television interview, which Professor Frank Zumbo of UNSW said should have been investigated by Australia's competition watchdog, AGL's chief executive, Michael Fraser, said: 'Historically gas has traded around $4 a gigajoule … I'm absolutely convinced that we're heading to a price regime of $6 to $8 a gigajoule … with all of the debate that's around coal-seam gas … that is simply going to put further upward pressure on prices.' About the same time, BHP Petroleum's chief executive, Michael Yeager, sent the

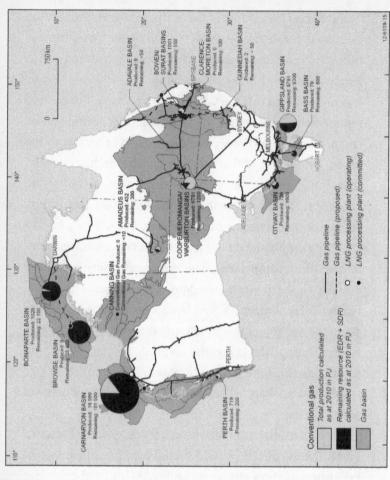

Map 4. Conventional gas
Proven reserves and prospective basins for gas found in sandstone reservoirs. This gas can be readily extracted with a small number of wells.

© Commonwealth of Australia (Geoscience Australia) 2012.

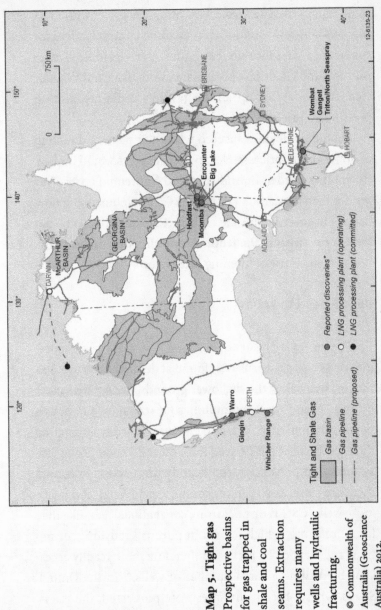

Map 5. Tight gas
Prospective basins for gas trapped in shale and coal seams. Extraction requires many wells and hydraulic fracturing.

© Commonwealth of Australia (Geoscience Australia) 2012.

Tight and Shale Gas

- Gas basin
- —— Gas pipeline
- ----- Gas pipeline (proposed)
- ● Reported discoveries*
- ○ LNG processing plant (operating)
- ● LNG processing plant (committed)

McARTHUR BASIN
GEORGINA BASIN

DARWIN
PERTH
ADELAIDE
MELBOURNE
SYDNEY
BRISBANE
HOBART

Whicher Range
Gingin
Warro
Moomba
Holdfast
Encounter
Big Lake
Wombat
Gangell
Triton/North Seaspray

750 km

same signal when he linked Bass Strait gas prices with LNG markets.[21] A few months later, as some contract prices for gas headed well into double figures per gigajoule, Fraser said that the Gladstone LNG plants were like giant vacuum cleaners 'hoovering up' all the gas they could get, putting pressure on prices.

Should companies succeed with their plans, a doubling of domestic gas prices driven by CSG exports would impose a crippling burden on industry and household costs – far greater than the carbon tax and the GST combined. But now that some governments are curbing CSG developments, the industry is twisting the truth by claiming that these restrictions are the cause of looming price rises.

WHEN WATER AND GAS MIX

Coal-seam gas is so named because it is methane gas trapped in porous coal seams, usually found at depths of 300 to 1,000 metres, well below the shallower groundwater aquifers used by farmers and rural communities for irrigation and household consumption. CSG is also known as 'unconventional' or 'tight' gas because it is derived from more complex geological systems that prevent or significantly limit access to it, and therefore extraction requires complex technology and many more wells. CSG is distinct from 'conventional' gas, like that found offshore, which is held in porous sandstone formations and trapped by rock, and therefore more readily accessible by drilling a small number of wells.[22] In the United States, where these techniques were pioneered, the gas is extracted mainly from shale beds and is therefore known as shale gas.

They say that oil and water don't mix, but in the case of CSG the gas is kept in place by the pressure of groundwater, which is why extracting it involves bringing vast amounts of water to the surface. If the coal seams are connected to aquifers used for drinking water and irrigation, then the release of water from coal seams could in turn reduce the amount of water in these vital aquifers. This potential damage is what makes CSG production particularly problematic for Australia. It involves messing with the Great Artesian Basin (GAB), the sequence of aquifers that stretches underneath most of Queensland west of the Great Dividing Range, into New South Wales as far south as Dubbo, across one corner of the Northern Territory and past the South Australian town of Coober Pedy.

After its discovery in the mid-1800s, the GAB spurred the development of inland Australia by providing fresh water for communities and livestock. Given the highly variable nature of rainfall in these parts, the GAB became the foundation for industries across an area of more than 1.7 million square kilometres of land, ranging from intensive cultivation towards the Divide to sprawling cattle stations further inland. Water moves through the porous gravel and rock layers of the GAB at a glacial pace, which means that any sudden changes to its usage may take decades, if not centuries, to rectify. This is why water extraction from the GAB is subject to very strict controls.

The water in coal seams typically has a high salt content

and contains traces of a range of chemicals and metals. It is an ancient resource that scientists refer to as 'dinosaur water', which is why bringing it to the surface raises serious issues about management and disposal. The amount of water extracted by CSG production varies from a few thousand to 100,000 litres a day. The production wells operating in Queensland have been producing 20,000 litres a day, although this usually tapers off over time.

When these figures are applied to 20,000 or even 40,000 wells over a period of decades, they add up to an awful lot of water.[23] Estimates by the National Water Commission, a federal–state advisory agency, predict that all the proposed CSG projects combined will increase extraction from the GAB by 300 gigalitres, or 55 per cent above the 540 giga-litres currently drawn from it by all users. Other estimates produced by the WaterGroup, a division within the federal environment department, indicate that the industry's con-sumption could be three to five times that amount.[24]

Even if the lower estimate applies, the implications are significant in two ways. First, companies don't yet have a viable plan for managing such a vast amount of brackish dinosaur water and all of the salt and contaminants that it contains. Second, as we'll see in the following sections, any connection between freshwater aquifers and coal seams risks serious damage to these precious resources in regional and remote Australia. These water resources could be damaged by methane gas, fracking chemicals and saline water leaking into aquifers.

FOR WHOM THE WELL TOLLS

The extent to which coal seams and aquifers are connected is an absolutely pivotal element in understanding the impact of CSG projects. The companies argue that they can prove categorically that there are no connections, which would mean it is safe to draw huge amounts of water from coal seams and then re-deposit waste water, but the reality is that there's great uncertainty about what lies deep beneath the surface. The available evidence indicates that no CSG well or coal seam can be seen as an island, completely isolated from the surrounding geology.

The federal environment minister, Tony Burke, says his approvals require companies to carry out thorough tests, thereby eliminating the risk of adverse consequences:

> The approvals that I've put in place say you have to test every individual aquifer. You test it seam by seam. And for each one, if it's water-tight and the companies are right, then you don't have an impact on groundwater. If it's not water-tight, if it's in fact porous and it is connected to the groundwater system or the Great Artesian Basin, then in those instances you either have to re-pressurise or re-inject the water after you've taken the gas out.[25]

This sounds very reassuring, yet a briefing paper that Burke would (or should) have read says there can be no such certainty. The 2010 paper by Professor Chris Moran and Dr Sue Vink, commissioned by Burke's own department, assessed the likely impact of CSG projects and said that there was a real risk of subsidence leading to the fracturing of the relationship between coal seams and aquifers. While all project proponents saw this as a minor risk, Moran and Vink believed that subsidence could 'change or cause fracturing in aquifers which may alter hydraulic connectivity'. The authors revealed: 'No proponents have considered the effect of faulting or fractures in their models.'[26] The danger of subsidence caused by CSG operations was also raised by the federal environment department in advice to Burke in 2010 for his approval of APLNG's 10,000-well CSG project: 'The proposals will also affect a significant number of stock and domestic users who rely on groundwater in the region, and have the potential to cause widespread subsidence on the land surface.'[27]

Graham Clapham, a Western Downs farmer who is faced with a mammoth Shell–PetroChina project on his alluvial farmland, says a group of local irrigators have proven that the coal seam below their land is in fact 'incised into' their aquifer, known as the Condamine Alluvium. Clapham is part of Central Downs Irrigators Limited, which represents farmers in the region who are fighting the proposed project. He is also one of several farmers who have taken the Queensland government to the Land and Environment Court over its granting of an exploration permit to Arrow Energy, the name for the Shell–PetroChina venture.

When interviewed at his farm for this book, Clapham had just returned from harvesting his corn crop, which had yielded an incredible 13 tonnes to the hectare. His farm is situated on a flood plain that has created extremely fertile soil known as 'black cracking clay'. The soil is three metres deep on his property. When it dries, cracks as wide as 15 centimetres appear, allowing it to 'self-mulch'. When watered, the soil expands to store enough water to grow an entire crop.

Clapham says the state government and the industry have steadfastly denied a connection between the Condamine Alluvium and the coal seams, known as the Walloon Coal Measures, but state government officials quietly urged Clapham's group to seek advice from John Hillier, an expert consultant and a former senior hydrologist with the Queensland government. Hillier was able to demonstrate that drawing water from the coal seams would drain the aquifer above it, although he could not give a precise estimate of the rate of flow.[28]

The experience of the Lloyd family and many other farmers living with CSG wells on their land indicates that water extraction from coal seams is already reducing the depth of freshwater bores. This anecdotal evidence was confirmed by an extensive May 2012 study by the Queensland Water Commission (QWC), which concluded that 528 bores in the Surat Basin, just north of the Darling Downs, would be affected in the long-term, while 85 would drop by up to a metre within three years.[29]

The QWC study confirmed that coal seams and aquifers are indeed connected. The QWC is a statutory water supply authority tasked for southeast Queensland. In response to community pressure it was given an ad hoc role in examining CSG impacts on groundwater, thus confirming that the former Labor government had embarked on developing this new industry without having a robust and independent regulator to protect water resources. The QWC concluded that CSG extraction is a risk when there are large numbers of fields near each other: 'If there are multiple well fields adjacent to each other, the impacts of water extraction from the fields on water levels will overlap. In these situations, a cumulative approach is required for the assessment and management of water level impacts.' It added: 'Therefore, all adjacent geologic formations are connected to each other. It is the degree of interconnectivity that varies.'[30]

Studies in the United States have identified instances of drinking-water contamination linked to shale gas production. A 2011 study by four academics from Duke University found that methane levels in drinking water were about 20 times higher in areas with gas production than the levels in areas with similar geology that did not have gas wells. The levels were high enough to create 'a potential explosion hazard'. The findings were published in the journal of the US National Academy of Sciences.[31]

WATER AND SALT

CSG production requires drilling through the aquifers of the GAB and then pumping a combination of gas and typically saline water to the surface, where it is separated. The water contains sodium chloride, sodium bicarbonate and what CSIRO refers to euphemistically as 'traces of other compounds', meaning a raft of chemicals and metals. The biggest challenge for industry and government is how to manage and dispose of these massive volumes of water, salt and toxic compounds. Based on the National Water Commission's estimates, CSG projects in Queensland alone will produce about 30 million tonnes of salt over the next three decades, while the industry's lower estimate of water extraction still yields 20 million tonnes of salt. In yellow truck terms, that's enough salt to fill 67,000 of them, which if lined up would stretch for 1,000 kilometres. Companies are exploring a range of solutions to dispose of it, from the building of enormous dumps to the re-injection of a brine solution underground.

Incredibly, most companies don't have a comprehensive solution in place for managing their waste water, yet they are going ahead with their projects regardless. CSG companies had been building massive evaporation ponds lined with builders' plastic – the sort used for cementing – but this practice was banned by the Queensland government in 2010 because of fears that the ponds could leach saline or toxic water in the event of a major flood. New South Wales has also banned these ponds.

Anne Lenz, the general manager for energy resources in the Queensland environment department (renamed as the Department of Environment and Heritage Protection after the 2012 election), said the ban is not universal, but applies 'in all but exceptional circumstances'.[32] Lenz added:

> If the dam proposed is an evaporation dam, the CSG operator must demonstrate that there is no feasible

alternative to managing the CSG water. An example of
this may be for low-scale exploration activities, where a
small amount of water is produced over a short period
of time, such as a few months.

This exception might explain why locals in the Chinchilla
region of Queensland, the epicentre of the state's CSG expansion, say that evaporation ponds still operate all over their
gas lands.

Without storage ponds, CSG producers have two options:
process the water in plants that use reverse osmosis, or re-
inject the waste water back into coal seams. In a 2010 policy
statement, the Queensland government said that 'injection
must be investigated as the first-preference water management technique, wherever feasible'. A new draft policy for
CSG water management, released in late 2011, outlines two
options: injection back underground, and processing so that
the waste water can be re-used. It is a fact that an enormous
industry is being rolled out in southeast Queensland without
a solution for dealing with a hazardous by-product. CSIRO
believes that this super-salty solution can be safely disposed
of in 'deep geological formations', but it concedes that there
are risks involved, because 'impacts on those aquifers need
to be considered'.[33] The Moran and Vink paper says that 'a
significant amount of further technical work is required to
determine appropriate reinjection targets, timing and water
quality/treatment needs'.[34]

After removing the salt, CSG companies also have the
option of dumping processed waste water with high levels of

contaminants into rivers, after having been authorised to do so by the Queensland government. In June 2010, the APLNG project gained approval from the state government to dispose of 20 megalitres of waste water a day from its desalination plant containing levels of chemicals and heavy metals that may exceed guidelines considered safe for animals, plants and micro-organisms. The contaminants included metals like aluminium, boron, copper and cadmium, and compounds like chlorinated hydrocarbons.[35]

With so many unresolved management issues, the new Queensland government's main response to farmer and community unrest over CSG has been to create a new body called the Gasfields Land and Water Commission. Its mission is to manage the 'coexistence' of the industry, regional communities and landholders, according to the LNP election policy. Jeff Seeney, deputy premier and minister for state development, says the new commission won't have any powers to impose fines or other sanctions, but it can make recommendations for such measures to be increased or introduced. The government has pledged a modest budget of $2.5 million in 2012–13 for the Toowoomba-based agency, but with plans for a chairman and six full-time commissioners, there may not be much money left over for on-the-ground operations.[36]

Heading the new commission is John Cotter, who previously chaired the state's farm lobby, AgForce, which has been criticised by some farmers for failing to defend their interests against CSG and mining development. At the time of this appointment in April 2012, Cotter provoked a flood of letters, emails and phone calls to a rural newspaper after he said in a letter to the editor that issues of groundwater, land access, confidentiality and compensation had been 'addressed and resolved' by a CSG consultation group that he chaired. In an interview with *Queensland Country Life*'s Troy Rowling, Cotter rejected claims that he may be biased because his son, John, heads a land-access consultancy that works for the CSG industry. John Cotter Jr, however, had put out a press release a week before his father's appointment, lauding the new commission. Meanwhile, John Cotter Sr has clearly signalled the need for less rather than more regulation on the industry: 'If those regulations are delivering that environmental practice then that's what those companies have got to live with. But if regulation is there impeding their progress ... then it needs to be looked at.'[37]

GETTING FRACKED

The technique of hydraulic fracturing, or fracking, adds another highly uncertain element to what goes on below ground. As many as four in 10 wells will be blasted with a mix of water, chemicals and sand at very high pressure. The available technology allows the blasting to occur at pressures of up to 15,000 pounds per square inch. In 2011, CSG companies presented evidence to the Senate Rural Affairs Committee that about one million litres of water, combined with sand and chemicals, are needed for every frack, with 10,000 to 35,000 litres of this being chemicals. At the high end, that's about two backyard swimming pools of chemicals for every frack.[38] (And for its part, CSIRO puts the upper limit on water needed at 10 million litres.) The danger is that the freshwater aquifers situated above the coal seams will be contaminated and that the chemicals will remain in the coal seam.

Gas companies say they can safeguard against such contamination by ensuring wells are properly sealed and isolated from the aquifers they pass through, but such promises have not held true in the United States, where the industry is more advanced. In December 2011, the US Environmental Protection Agency (EPA) – the kind of robust institution that is sorely lacking in Australia – announced the draft findings of a study that found synthetic chemicals such as glycols and alcohols 'consistent with gas production and hydraulic-fracturing fluids' had been identified in drinking water near the town of Pavillion, Wyoming, where a Canadian gas company has sunk 150 gas wells. The levels identified were 'well above' the safe limits for drinking water. This revelation came after the EPA drilled two of its own wells into aquifers to test the quality of drinking water. The EPA statement noted the risk of chemicals left behind that could migrate into aquifers used for drinking water:

> Given the area's complex geology and the proximity of drinking water wells to ground water contamination, EPA is concerned about the movement of contaminants within the aquifer and the safety of drinking-water wells over time.[39]

CSIRO argues that fracking has been used in the oil and gas industry since the 1940s, and that the technical aspects are well understood, although it concedes euphemistically that 'environmental impacts are less well characterised'.[40] Guarding against contamination of aquifers involves lining wells in steel casings and then cementing them in place to 'isolate' them from the aquifers that they intersect. Before fracking is carried out, 'the integrity of the cement bond between the casing and rock needs to be confirmed and verified'. Note that in spelling out how the industry should operate, CSIRO consistently writes in the passive voice, thereby not putting responsibility on anyone in particular. CSIRO's commentary on what happens to the fluids used in fracking takes the prize for omission of relevant facts: 'Ultimately, 60 to 80 per cent of the fracking fluid flows back to the well from the coal seam. These fluids are brought to the surface inside the steel casing. This fluid is then pumped to lined containment pits or tanks.' In other words, 20 to 40 per cent of these chemicals remain underground, thereby posing a real threat to human health, and that of livestock and cultivation, should they flow into aquifers as they appear to have done in Wyoming.

The chemicals used in fracking are a major concern, as some of them have been banned overseas. CSIRO says that chemicals used by companies 'may be commercially confidential'. Its list of fracking chemicals includes: sodium hypochlorite (used in bleach and as a biocide in swimming pools); hydrochloric acid (a strong corrosive acid); surfactants (used in soaps); guar (used as a gelling agent); acetic acid (the basis of vinegar); and bactericides (to inhibit bacteria forming that may corrode the steel casing or plug the permeability in the fracture and coal seam).[41]

Australia's national chemical regulator, the National Industrial Chemical Notification and Assessment Scheme (NICNAS), is reported to have assessed only two out of the 23 compounds found in the authorisations given by state governments to CSG companies.[42] In answer to a series of questions for this book, NICNAS said it had examined a number of chemicals used in fracking, but could not say how many. It admitted that it had 'not undertaken public risk assessment of these chemicals'. However, NICNAS was 'working urgently' with other Australian government agencies to formulate 'a systematic approach' for assessing the risks to human health from fracking, but it did not say when this work would be completed. Perhaps this regulator can't be blamed for its shortcomings, as it's far from the robust institution that seems to be needed in this area. In 2011, NICNAS received just $300,000 in federal funding, and most of its $9.3 million budget is financed by levies on industry.

One of the problems facing the regulator is that information from industry is patchy. NICNAS said there is 'inadequate' data on the volume and use of these chemicals by gas companies. 'This information will be required to fully assess the health and environmental impacts of the chemicals. NICNAS is liaising with CSG industry operators to determine this information,' NICNAS said, without giving any idea of how and when this information might be obtained. As Australia launches itself into a new chemically-intensive industry, this is hardly a reassuring state of affairs.

But this isn't the end of the fracking story. The petroleum industry in the United States is developing technology that will enable it to conduct what it calls 'super fracking', as well as horizontal fracking. A decade ago, virtually all of America's natural-gas production came from traditional gas or oil wells reached by vertical drills. Companies have learnt how to drill horizontal wells that, combined with fracking, have broken up shale formations to release vast amounts of gas. In 2009, about 14 per cent of America's natural gas came from shale, but this is projected to increase to almost half by 2035. High prices have encouraged even greater innovation, to the point where oil services companies are finding ways to create longer, deeper cracks in the earth to release more oil and gas.

The leading fracker in the United States is Halliburton, the company whose cement work on oil wells was linked to both the 2009 Montara and 2010 Gulf of Mexico oil spills. It has now produced a technique called RapidFrac. Its competitor Baker Hughes has developed technology to create 'super cracks' by blasting deeper into dense rock to create wider channels and disintegrating balls, which turn into powder 'like an Alka-Seltzer' after a couple of days. Finally, the oil service firm Schlumberger has developed a technique called HiWAY, which it claims yields larger production of oil and gas while using less water, sand and fewer chemicals, thereby reducing costs.[43]

The United States has been involved in this industry for much longer than Australia, and on a vastly greater scale; it has clearly developed more expertise in how to manage these projects. Even so, the long-term impacts of extracting tight gas, and techniques like fracking, are still not well known, and the United States has had more than its fair share of problems.

The risk for Australia is that this technology is likely to fall into the hands of inexperienced operators. Roy Michie, who worked for the fracking arm of Halliburton in South Australia's Cooper Basin and Queensland's Surat Basin, says the industry is dominated by 'cowboys' and governed by substandard regulations, especially when compared to the rigorous standards of the underground mines where he had previously worked.

'Oil and gas just doesn't seem to have any rules,' he says, 'it just depends on who is running the show on the day as to what you will do and how you will do it.' He added that the industry is using 'very nasty chemicals'.[44]

SPIDERWEBS

Extracting the tight gas trapped in coal seams is difficult in that it requires many more wells than conventional gas extraction. These wells have to be accessed by pipelines and roads, producing a highly invasive network that will have an enormous impact on farmers and rural communities. So far, the developments in Queensland indicate that each well generates about one kilometre of roads, so the wells destined for Queensland's productive farmland could generate 40,000 kilometres of pipelines and access roads. Such a network is likely to harm wildlife and diminish farm productivity, thereby threatening the viability of food producers that we and the rest of the world depend on. Advice from the federal environment department to the environment minister, Tony Burke, for the APLNG project, the biggest approved so far, says of the impact of the roads and related infrastructure: 'It is not possible to identify detailed environmental impacts with precision.'[45]

The road network is built to facilitate frequent access to the wells by maintenance crews, but this means a large number of vehicle movements that raises the risk of spreading weeds and pests. In the event of a disease outbreak like foot and mouth – heaven forbid – such vehicles' movements could spread the disease. Or, if a region had to be quarantined because of such an outbreak, CSG companies might find themselves in the invidious position of being unable to carry out the required maintenance on their wells. The Queensland government has imposed explicit requirements on CSG companies to minimise the spread of pests and weeds by washing down their vehicles. Its Land Access Code states that companies must ensure that each person acting on its behalf 'washes down vehicles and machinery before entering a landholder's land', and that records of such wash-downs must be maintained.[46] But some farmers in the gas fields of the Darling Downs keep cameras on their gates, and they report regular entry of muddy CSG vehicles into their properties, even though locals say there was an outbreak of the noxious weed parthenium at APLNG's plant in 2009.

The pipeline network also carries significant risks for the communities living nearby, and for the environment as a whole. Wells have blown up in Australia, including one managed by Arrow Energy, which sent a shaft of gas and salty water 90 metres into the air in mid-2011. There have been numerous reports of leakages from wells and pipelines. And in fact, pipelines have pressure valves designed to leak gas and water – the gas vents placed along the CSG pipelines in Australia attest to that – although this undercuts the industry's claim to produce a clean green energy source. One peer-reviewed study by researchers from Cornell University, published in the respected journal *Climatic Change* in April 2011, argues that leakages from shale gas pipelines in the United States have made gas production even more greenhouse-intensive than oil and coal. The industry has tried to discredit the study, but there is no doubt that there's a lot of leakage.[47]

The spiderweb of wells, vents, access roads and pressurisation stations is already undermining farm productivity. Broadacre farming requires long clear runs and these will be inhibited by having numerous wells, vents and roads dotting and criss-crossing cultivation land. The experience of the Lloyds and their Tara farm shows how good cultivation land is already being lost to CSG infrastructure, and this is still in the early days of expansion, when companies have been attempting to reduce their impact by putting wells on grazing land whenever possible. As the expansion unfolds, more cultivation land will join the spiderweb. As a Condamine farmer explained, the patchwork of blocks that can been seen from the air has been designed this way for a very good reason. The patchwork allows farmers to work their fields with long straight runs of two or three kilometres. But once pipelines are running across the patchwork, farm machinery won't be able to work the land in the same way. CSG companies in the Chinchilla region have used legal pressure to force farms to put their main gas pipeline, which measures

42 inches in diameter, right across the cultivation blocks. The affected farmers complain bitterly about their treatment to their mates in private, but strict confidentiality prevents them from speaking publicly.

The companies are responding to farmer resistance, such as that of the 'Lock the Gate' movement, by buying up vast tracts of farmland. A Condamine farmer who asked not to be named said that within a 16-kilometre radius of his farm, CSG companies had bought up half the acreage, about 8,000 hectares of intensive cultivation land. One of the farms was owned by a man in his eighties who now lives in town, but is often seen driving back to the boundary fence to look over land that had been in his family for generations. A neighbour says the elderly farmer sold after the company wore him down and then made him a decent offer. The land is now being used to recycle waste water from a CSG operation. 'He didn't want to live with gas companies. Every day you have contact with them. They are a total intrusion. This is what gets to people more than anything else,' says the neighbour. Origin Energy confirms that its APLNG joint venture has acquired 65,000 hectares of farmland across Queensland. Some of this land is irrigated with treated waste water from its CSG project, and the company even runs 5,000 head of cattle.[48] QGC has land holdings that cover 37,000 hectares of farmland in Queensland.[49]

The impact of this development is manifestly taking its toll on the quality of life of farmers, causing some – especially the older generation – to sell up and leave for good. Darling Downs farmers Megan and David Baker have Peabody's

Wilkie Creek coal mine just 20 metres from their boundary fence. For the past 16 years, it has stirred up a constant barrage of noise and dust. They now have to contend with Arrow Energy's nine CSG wells that bring a steady stream of gas workers onto their farm. Gone are the days when the Bakers felt comfortable skinny-dipping in their dam. 'Privacy is non-existent,' Megan says.[50] On most evenings, the Bakers are interupted by a phone call from Arrow, Peabody or one their contractors.

Neither the coal nor the CSG gas company has needed to buy the Baker farm in order to operate. But dealing with both companies has affected the Bakers; David Baker has cancer, and the stress of constant and complex interactions with both a mining company and a gas company is not making his recovery any easier. The strain seems immense. Says Megan: 'For me, it's a day-to-day drag. It's just hard emotionally. It's taken over my life trying to manage it. The paperwork, the impacts, keeping up with it, knowing our rights.' But what makes life even worse for the Bakers is that they feel they are stuck where they are; they don't think anyone would want to buy a farm located so close to a coal mine, with CSG wells as an added disincentive.

These David and Goliath battles are likely to be ongoing because the CSG expansion underway in Queensland is locked in by the massive investment in three LNG plants at Gladstone, which are now being built. Once completed, these plants and related export contracts will create demand for an ever-expanding supply from the gas fields of the Darling Downs and the Surat Basin.

But farmer resistance could mean that the gas companies don't secure enough supply to fulfill their contracts and make these plants pay their way. The gas companies would have to source that supply from elsewhere in the domestic market, putting added pressure on local gas prices. Supply could also be restricted by the prospect of stronger regulation as a new state government in Queensland, along with parliamentary inquiries by the NSW Parliament and the Senate, respond to community demands for a more robust and transparent regulatory regime to ensure that the Darling Downs and the Liverpool Plains don't succumb to the same fate as farming regions affected by earlier booms. One way or another, Australia's fast-tracked journey into gas land is likely to end in tears for many.

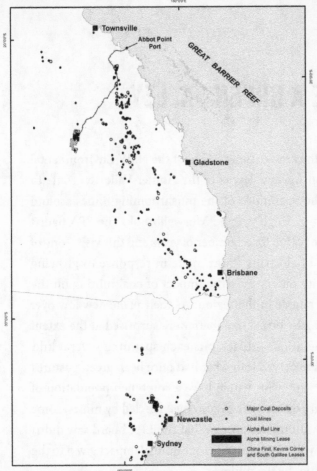

Map 6. Mines and mega-mines

Existing and prospective coal mines now dominate the prime farming regions of the Hunter Valley northwest of Newcastle, the Darling Downs west of Brisbane, and the Bowen Basin west of Gladstone. These will be dwarfed by the proposed mega-mines in the Galilee Basin further inland from Gladstone, which will generate thousands of shipments each year through the Great Barrier Reef.

Source: Geoscience Australia Datasets – Mine and resource locations.

ONCE A GREEN VALLEY

In 2009, the now-defunct board of the NSW Environmental Protection Agency flew over the Hunter Valley en route to inspect the operations of the mushrooming mines around the towns of Singleton and Muswellbrook. The EPA hadn't visited the region for a number of years and this visit – one of its annual field trips – was made in response to growing community outcry over the impact of coal mining on the health of people in the region. As their plane flew low over the mines, the board members were surprised at the extent of the operations – black pits each spanning several kilometres were carved into what had once been green pastures. And the two towns, which have a combined population of more than 30,000, were becoming encircled by mines, some just a few kilometres away. What the EPA board saw didn't match up with the assurances of minimal impact given to the communities by the miners.

On the day of the visit, the industry appeared to have gone into hibernation. Most of the 13 drag lines that operate in the valley were not working, but even so the visitors could see the dust reaching a height of about 800 metres. As they approached Singleton, they saw two adjoining pits near Mt Thorley (the mountain itself having long been removed) measuring more than 13 kilometres long. At Muswellbrook, they saw the Mt Arthur North coal mine, four kilometres south of the town, which is seven kilometres long and three kilometres wide. Next to it were two other large pits for the Bayswater and Drayton coal mines. Newer mines were expanding immediately to the west and east of the town, one as close as one kilometre from the town's edge.

After returning to Sydney, the EPA board summoned the heads of two government departments and asked for an explanation of what was happening in the Hunter Valley. They asked about the extent of rehabilitation work and compliance with the conditions imposed by the government when it approved the mines, but received surprisingly few answers. 'They couldn't answer any of our questions about how much land had been rehabilitated. The departments couldn't say where the biodiversity offsets were. No-one was monitoring to see if the mines were doing what they say they are going to do,' recalls Bev Smiles, a Hunter Valley landowner who was appointed to the board as a community representative.[51]

The Hunter Valley is home to about 30 coal mines and produces about 100 million tonnes of saleable coal a year, or a quarter of the nation's production. Between the towns of Singleton and Muswellbrook is Australia's most intensive mining region, where more than 20 mines operate in an area that measures just 25 by 60 kilometres. Mining has made the area within this cluster a virtual no man's land where very few people now live. The total workforce in the Hunter mines is about 14,000 full-time employees, including contractors, representing 15 per cent of the region's workforce. Two decades ago, the region still had a balanced economy that included tourism, wine, thoroughbreds and defence, but all of this began to change in the 1990s, when the state government adopted an open slather approach to the approval of big new mines, while allowing existing ones to expand. Mining has come to dominate the region to the point where it is squeezing out other industries, most notably wine production and tourism. Communities like Camberwell and Wambo have been virtually erased; others like Bulga and Broke are staring down the barrel of new expansion plans that will almost certainly wipe them off the map. Those who have remained have been driven to distraction by the dust, noise and blasting.

In the relentless drive to develop new mines, pioneering farming families, like Wendy Bowman's, have been pushed aside. The loss of water as a result of mining has directly undercut the ability of agriculture to operate. Just a few years after the Ravensworth underground coal mine began operating in 1987, the miners drilled into the aquifer and Bowman and other neighbours found that their bores suddenly went dry. In the early days, this risk wasn't considered by the government when it approved new mines. 'No-one in the early days was talking about water. There was no thought about underground water,' she says.[52] Bowman, 78, who has become an anti-mining activist, says the mine showed little regard for the impact on neighbours: 'They started mining on their fence line, near the dairy cottages, so the men could not sleep at night. The dust was unbelievable. When they blasted, you could not see.'[53] Eventually, the mine gained approval to evict her.

Saddest of all for Bowman was the loss of her family home, originally built in 1861. It had cedar doors and window frames, hand-blown glass windows, and was built with hand-made bricks. When the government wanted to widen the main road in the early 1900s, the family dismantled it and rebuilt, brick by brick, 400 metres up the hill. But when the Ravensworth mine acquired the property, it gave Bowman just 21 days to leave, and then bulldozed the house. Bowman was a widow living on her own, and she cried herself to sleep every night for months after leaving her property.

The winegrower Brett Keeping, who runs the Two Rivers vineyard near Denman, says encroaching mines in the valley have squeezed the non-mining sector by driving up costs and the Australian dollar. The mines have even presented a physical constraint on the ability of the region to attract tourists – miners and contractors have booked out all of the tourist accommodation, making it well-nigh impossible to attract visitors.[54] Mines have bought out some vineyards, while others have had mines open up next door or across the valley. Keeping faces the threat of having a mine open up about four kilometres from his winery. He dreads the thought of visitors looking across at an open-cut coal mine as they have lunch on his terrace. He says the big problem for the winemakers is that they suffered a knock-out blow when they first tried to resist in the mid-1990s; the vineyards 'lost the fight first up'. It was an epic struggle in which the ruthless pursuit of development by state governments prevailed.

The experience of the Hunter Valley contains powerful lessons for regions like the Liverpool Plains and the Darling Downs, which are trying to resist the expansion plans of the coal mining and CSG companies.

A TALE OF TWO BOBS

The shrewd and tenacious businessman Bob Oatley, now in his eighties, is best known for his ownership of the *Wild Oats XI* maxi-yacht and a luxury resort at Hamilton Island, and his creation of Rosemount Estate wines, which first produced commercial vintages over 40 years ago. He is less well known for his fight to stem the spread of mining projects in the Hunter Valley, largely because he was beaten, unfairly so. When the state government authorised new mines in the mid-1990s, Oatley believed that the industry was on course to dominate the region. As a result, although in his sixties, Oatley fought with courage and serious money. He would have won had it not been for another Bob – the former NSW Premier Bob Carr – who has since garnered a profile as an environmentalist.

In the mid-1990s, Rosemount Estate was one of the most successful wine labels in Australia, at the vanguard of Australia's export push. Rosemount's wine-making plant in the Hunter Valley was a formidable operation that employed more than 300 people. But when approval was given for the Bengalla open-cut mine to go ahead – located on verdant pasture just 2.5 kilometres west of Muswellbrook – Oatley believed that this marked the beginning of the end for the wine industry in the region. Oatley challenged in the Land and Environment Court and won. The court ruled that the Carr government's original approval for Bengalla was invalid and that its actions were unsound and unreasonable.[55]

The response from the Carr government was swift and brutal – it rushed legislation into parliament to override the court's decision and approve the mine. Known as the *State Environmental Planning Policy 45* (SEPP45), this planning instrument gave the government extraordinary power to approve new mines where mining had hitherto been prohibited. Justice Paul Stein of the Land and Environment Court said the effect of the new policy was plain: to 'make permissible mining which will have, or is likely to have, adverse and environmental effects in environmental protection zones'.[56]

The *Sydney Morning Herald* said in an editorial that there was an '*Alice in Wonderland* quality' about the NSW government's involvement in the controversial mine. Not only did the bill seek to override a court decision, 'it also seeks to pre-empt a decision by the Court of Appeal – an appeal brought by the Government itself and a decision that has not yet been delivered'.[57] The newspaper rightly foreshadowed the role of the special law in opening up the Hunter:

> The bill has wider application than just the proposed Bengalla mine. It also validates the Government's State Environmental Planning Policy 45. This is a curious planning instrument which gives [the minister, Craig] Knowles powers to override local environment plans and approve mines even where mining is prohibited.

Oatley had invested enormous amounts of money in trying to stop the mine, and even after the government put SEPP45 into law he took out full-page advertisements in newspapers pleading with the government to reverse its decision. In one advertisement he said the government was willing to 'ignore the existing planning code and abandon its electoral responsibilities to Rosemount Estate, its employees and the people of New South Wales'. Oatley said the people of New South Wales could have no confidence in the integrity of the state's planning codes: 'It is clear that you can alter policy at the stroke of a pen without any consideration or public participation and debate.' In a prediction that has largely been realised, Oatley warned that the move would give primacy to mining above every other consideration, be it the interests of other industries or the environment:

> It needs to be appreciated that the proposed legislation does far more than legitimise the Bengalla project. There are many areas of New South Wales where mining is only permissible if certain criteria are satisfied. The effect of this legislation will be to destroy this control mechanism and make mining the primary use for all land irrespective of any agricultural or other potential usage. It is extraordinary that legislation of such far-reaching consequences can be proposed without any opportunity for public discussion.[58]

Bengalla began operation in 1999 and is producing 5 million tonnes per year, but like so many other mines it is now to be expanded. The mine's major shareholder, the London-based multinational Rio Tinto, has plans to triple production to 15 million tonnes per year. Bengalla was soon followed by many other new mines, leading to a steady increase in production. Between 1999 and 2008, coal production in the Hunter and Newcastle regions has increased by 33 per cent to 141 million tonnes, yielding 105 million tonnes of saleable coal.[59] When burnt in power stations or steel mills here and overseas, the coal from the Hunter Valley produces a conservative 225 million tonnes of greenhouse gas emissions a year. That's more than the greenhouse gas emissions of the Philippines, ranked as the world's thirty-fifth biggest emitter and with a population of 94 million, and of Vietnam, population 90 million. After having aggressively driven this expansion, Bob Carr told the Senate in his first speech as foreign minister in March 2012 that climate change was the greatest threat to human civilisation.

After the mine went ahead, Oatley made plans to vacate the business and the region. He sold Rosemount to Southcorp in 2001, which was then acquired by Foster's in 2005. In 2009, BHP bought one of the last surviving vineyards of the former Rosemount empire. The 369-hectare Roxburgh vineyard had been part of a buffer zone between the existing Mt Arthur mine and Denman Road, but the acquisition also gave the company Roxburgh's 'high security' water allocation, worth 800 million litres a year from the nearby Hunter River. Brett Keeping says that Roxburgh was an iconic part of the Upper Hunter winemaking history, as it had produced some of Australia's best chardonnay. He adds that many other vineyards are being acquired by mining companies, including six bought by Xstrata's Mangoola (formerly Anvil Hill) coal mine.[60]

The Oatley family re-emerged with a new winery in the Mudgee region, located about 260 kilometres northwest of Sydney. For now, it might seem that mines are a long way away, but in fact plans are afoot to open up virgin coal provinces, including in the Bylong Valley, where the former NSW Labor powerbroker Eddie Obeid has bought a farm with a coal exploration lease over it. There are also plans for a silver and lead mine near the hamlet of Lue, population 800, where the former Supreme Court judge Morris 'Dusty' Ireland has retired. Ireland spoke at a rally in Mudgee in March 2012, saying that he was horrified at the thought of a lead-silver mine opening just 1,500 metres from a primary school. The mine is being developed by Kingsgate Consolidated, which describes the project as a silver mine even though most of the ore produced will be lead and zinc. The company's own figures indicate that 1.74 million tonnes of lead and 2.32 million tonnes of zinc will be produced over the life of the mine, compared with 30,700 tonnes of silver, should the company proceed. Kingsgate is like many small- to medium-sized mining companies that use Australia as a safe haven while extracting wealth from riskier countries. This set-up enables companies to leverage Australia's diplomatic clout in these places. While Kingsgate is listed on the local stockmarket and has an office in Martin Place, Sydney, most of its income over the past decade has been derived from a gold mine in Thailand, while it is also pursuing a silver and gold mine in central Chile.

PASTURES OF PLENTY

James Lonergan and his father, Jim, think their dairy farm near the mining town of Muswellbrook is just 'perfect', and the awards they've won for their milk support that claim. On 365 hectares of lush green river flat, they run 90 milking cows that produce 1,500 litres of milk every day of the year. They have a thoroughly reliable water supply to keep their pasture green all year round – they draw water from the Hunter River that flows along one boundary, and they also have bore water. 'This country is the best you will get for producing food,' says James Lonergan.[61]

The Lonergan farm, established in 1905, is ranked in the top 5 per cent of all dairy farms in Australia, based on its ability to produce milk with very low bacteria content (known as the somatic cell count). The average cell count for most milk is 250,000 to 300,000 parts per million – the Lonergans' is half that amount. The count reflects the cleanliness of their operation: the farm has been awarded the Dairy Farmers 'milk excellence' award every year since its inception in 2001.

The Lonergans have already seen the impact of a mine located about two kilometres from their boundary. In 2003, water in one of their wells, and all of the wells on a neighbour's farm, went dry overnight. The neighbouring Dartbrook long-wall mine, which began operations in 1994, had cracked an underground aquifer and flooded. After this, the Lonergans watched as a steady stream of trucks carted off water that had been pumped out of the mine. The owner, US giant Anglo American, took responsibility for the loss of this crucial resource and agreed to pump water to the Lonergans' dry well and those of the neighbour. James remembers his grandfather telling him how the family had always been able to rely on the well: 'Pa said it never went dry.' Jim Lonergan, 62, says with palpable regret that the water 'will never come back'.

Yet this setback seems trivial compared to what they're facing now. All around the Lonergans, dairy farms have disappeared as new mines have been approved. In the 1970s there were as many as 400 dairy farms in the Singleton Shire; now there are as few as 19. The Lonergans' is the last privately owned one in the district, as the others are leased from mining companies that bought up small blocks. The Bengalla mine is located just eight kilometres to the north of their farm, and when the southerly wind blows, they can smell the sulfur. Next to Bengalla is about 4,000 hectares of land acquired for the development of the ironically named Mt Pleasant open-cut mine. The project is also being developed by Rio Tinto, and the Lonergans fear that the multinational will get approval for another mega-mine. They also fear that, like many other underground mines, the nearby Dartbrook mine will go open-cut.

Given these prospects, James, who married last year, doesn't see himself raising children on these perfect pastures: 'I don't think there's a future here if they start. There's no point trying to stay beside them and get all the dust, and the water.' James fears for the quality of the water supply if Mt Pleasant goes ahead. He points to the shape of the land and explains how the runoff will flow into the Hunter River, just above Muswellbrook.

The boom around the Hunter Valley is already putting the viability of non-mining businesses on the line, squeezing them out. The local refrigeration mechanic can't keep employees; service firms such as the engineering companies say bluntly that they are no longer interested in working with small businesses. Even the well-known engineering firm Theiss used to do work for the Lonergans, but no more. 'Now we have to do it ourselves,' says James.

As the mining expansion has grown larger and larger, the attitude of the mining companies has hardened towards the remaining farmers. In the early days of the Dartbrook mine, there were a number of environmental officers who dealt with the farmers in a frank way. One of them, who 'used to tell the farmers what was going on', was sacked. Others like him have also been sacked.

The Lonergans have sought to tell their story to the media, but so far no-one has published the details of their extraordinary situation. In 2011, the Nine Network's *60 Minutes* came to interview them about the dry wells, but then all went quiet. It seems the program may have been pressured by the mining companies. There's even less interest and support from the farmers' traditional political representatives. The Lonergans' local member, George Souris, who has represented the Upper Hunter electorate for 24 years, has 'taken the side of the miners', even though he is a National Party MP. 'The whole problem is they're not the party of the farmers,' says James. Perhaps it should not have come as a surprise, as Souris cheered when Bob Carr's legislation for open slather development went through the NSW Parliament in July 1996:

I congratulate Bengalla Mining Company Pty Limited on the way in which it has approached this development and engaged in property acquisitions and necessary community relations. I commiserate with Rosemount Estates Pty Limited – whose representatives are in the public gallery today – which has a genuine conviction about the issues it has raised. Rosemount is entitled to those convictions, to the full extent of all available processes and to its day in court. I sympathise with the fact that Rosemount is aggrieved that this legislation has had to be introduced. However, the Parliament is the supreme legislature and the imperatives of the development remain.[62]

After stepping down from the Land and Environment Court in the late 1990s, Justice Stein admitted that developers had become so influential that politicians and bureaucrats couldn't say no. There was a 'sense of virtually never saying "no" to a proposal, however irrational or environmentally damaging it might be. Controversial proposals are massaged, coerced or "mediated" through to a "yes" – what politicians like to call a "win-win" situation.'[63]

These words ring true in much of New South Wales. On alluvial flood plains that farmers thought would only be mined in their worst nightmares, mega open-cut mines have been authorised, and more are planned. The public servants in the state's development department egg on the politicians. When a Liverpool Plains farmer first learnt that a mining lease covered his property, and another was issued across the valley, he sought a meeting with the planning department to try to explain the uniqueness of this productive area and the damage that mining would do to its water systems. At this September 2006 meeting attended by George Souris, the Liverpool Plains farmer, Tim Duddy, was told in a bullying tone by a senior public servant: 'We are going to build a mine on the Breeza Plain and there's nothing you can do about it.' The official was happy to see a productive valley with alluvial aquifers turned into open-cut pits, just like in the Hunter Valley.

PLASTICINE AND POST-IT NOTES

The residents of the towns of Broke and Bulga near Singleton have lived close to some of the biggest mines in the region for the past 30 years. After such a long period, many residents hoped the mines would soon be nearing the end of their lives. But the surging price of coal has encouraged companies to dig deeper, leading ambitious miners to engage creatively with the local communities.

The businesswoman and agriculturalist Mary Tolson has lived beside Xstrata's Bulga coal mine for three decades, after moving to the area at the age of 18. Tolson and her family have lived as close as one kilometre from the mine at times, and for the past decade, 2.5 kilometres from the lease boundary. It was in this environment that Tolson and her husband, Trevor, built a farm and then a rural services business, and raised two children. 'It is like living near a freeway,' she says. 'You hear the acceleration, the deceleration, the coal being dumped into dump trucks. I have got it at my door.'[64]

When the mine started in 1982, it took Tolson and her neighbours five years to convince the then Saxonvale mine to change to quieter dump trucks. In a recent complaint to Xstrata, Tolson said its blasting had caused cracking to her floor tiles, plaster cornices and patio cement, resulting in 'time-consuming, costly and stressful administrative battles with Xstrata to have the damage acknowledged, let alone rectified'.[65] Following dozens of noise complaints, Tolson has not been able to get any noise data from the company, which says it is still compiling 'historic data'.

Over the decades, the NSW government has received a steady stream of applications to expand the mines near Bulga, and in all instances it has agreed, notwithstanding the toll on residents. The latest batch of applications includes a proposal from Rio Tinto to expand its combined Warkworth–Mt Thorley operation to extract another 200 million tonnes of coal. The mine will expand west to the point where it will destroy part of Bulga, a town of 320 people, consuming 764 hectares of woodland and 141 hectares of a 'permanent' conservation zone that was announced with some fanfare by Premier Bob Carr in 2003 when the mine last got permission to expand. The latest expansion will take out all of Saddle Ridge, which protects Bulga from the mines, while 24 homes are up for acquisition by the mine and an eight-kilometre stretch of the Wallaby Scrub Road. The permanent conservation zone was part of a legal agreement between the state government and Rio Tinto, which states that the company must rezone the land and prevent further mining, something that Rio conveniently failed to do.

To review this contentious proposal, the NSW government asked the Planning Assessment Commission (PAC), a special review body, to set up a panel that in this case was chaired by Dr Neil Shepherd, a former NSW departmental head and a former head of the EPA. The panel also included Gabrielle Kibble, who was head of the Department of Planning from 1995 to 1997 – exactly the time when the Bengalla legislation was rushed through state parliament.[66] (Farmers who are contesting Nathan Tinkler's Maules Creek coal mine in northern New South Wales also have to contend with Kibble as chair of the PAC for this project.)

The Bulga commission cited a raft of problems for the people of the town as a result of the road closure and the impact on property values, and it foreshadowed that 'loss of community identity may result from the departure of existing residents as they choose to move away from the expanding mine complex'. But such costs were greatly outweighed by the prospect of even greater coal production, especially at a time of high prices. It seems nothing can offset the perceived benefits of development. The commission wrote:

A number of rural communities have been faced with this situation in the past. In almost all cases the mines have been approved and the communities have either been radically altered in character or become non-viable. With the current price of coal this outcome is almost inevitable when the overall economic benefits of the mines are balanced against the local community impacts.[67]

To locals in Bulga, the state government's willingness to allow Rio Tinto to breach its 2003 agreement simply proves that people living in the shadow of mines don't matter at all. One long-time resident, Ian Hedley, whose family has lived in the area since 1821, says the power and influence of the miners has grown to the point where they are simply answerable to no-one. Even though his mining equipment business has prospered as a result of the boom in the region, he thinks that Bulga's betrayal underscores the excessive power that the industry now wields. 'When they first came here they were fantastic to deal with. Now there's no effort to rein them in,' he says. He predicts that the expansion will make the town unliveable.[68] He believes that the mine's superficial approach to dust and noise reduction is aimed at making Bulga so uncomfortable that people will leave and Rio will be able to 'drill right through there to the national park'.

The Warkworth operation isn't the end of the onslaught facing Bulga. Xstrata also has ambitious plans there. In late 2011, Mary Tolson received a visit from the company's community liaison manager, Ralph Northey, and two other Xstrata executives. At the meeting, Northey unveiled what the company calls its 'optimisation' strategy for the Bulga mine. It plans to dig twice the existing depth and create an eight kilometre–long pile of overburden that would close a road, forcing residents to drive several kilometres around the mine to reach Singleton. It would take three and a half years to remove the overburden and the 'optimised' mine would be in operation for another 23 years.

In this early phase, Xstrata's strategy for obtaining approval involves demonstrating to the government that it has conducted community consultation. To this end, Xstrata hired a consulting firm to stage meetings to discuss the community's 'vision' for the future. The company described its 'Visioning Project' as a way 'to better understand and support local communities' aspirations for the future ... to bring together community views of the future and create momentum to realise this vision'.[69]

At a November 2011 meeting targeting wine, accommodation and tourism operators, about 20 residents sat at tables stacked with large aerial photos of the three local villages and packets of post-it notes. The consultants asked the participants to write on the notes the things that they valued in the community and to place each note on the relevant area. Some participants took to the task with gusto and wrote 'No Noise' and 'No Dust' and the like on the notes and put them on the map. Next, the consultants put lumps of plasticine on the tables and asked the participants to make models of things that symbolised their 'vision' of the future. At this point, Tolson became frustrated and angry. She told the consultants: 'I have come here for a workshop. I have not come here to play with plasticine. My vision is that Xstrata clears out of here and does not go ahead.'

What most angers Tolson and other Bulga residents is that Xstrata is planning to expand the mine while being well behind on its rehabilitation work. After two decades of mining there are huge expanses of overburden waiting to be rehabilitated, while watercourses have not been restored.

When residents raised this, Ralph Northey apologised and said that the government had not been 'pushing' the company or industry to do it. 'There has been no push,' he said, while promising to do better in the future. Northey declined later to comment on these remarks.[70] In a reply to Tolson, Xstrata said it 'acknowledges the community's concerns about the delay in rehabilitation of the overburden spoil near Charlton Road', and that it would rehabilitate the most visible portion in 2012.[71] In turn, Tolson wrote that she and her family had been 'celebrating' the prospect of the mine closing only for the company to announce that mining would be 'optimised' until 2035. She said the mine would operate 'well into my retirement years or till my death', and regarded this as a 'double life sentence' that had destroyed the vision she had for her future when she moved to Bulga in 1979.[72] She told the company that she and her family were 'outraged, distraught and depressed'.

Tolson's experience shows how companies avoid responsibility for the impact on people living in their shadow. Bev Smiles, the former community representative on the EPA board, knows of dozens of families like the Tolsons who have lived on the edge of mines and been tormented by the constant assault of noise and dust, and yet received no compensation. Companies only have a responsibility to properties named in the government's approval as 'affected'. But the government doesn't decide who is deemed to be affected, says Smiles: this is derived from the analysis in the company's EIS. The government generally accepts at face value what the company tells it, and copies this list of affected properties in the approval conditions. In the case of Mary Tolson, Xstrata didn't even respond to her request, in a formal complaint made in November 2011, to be bought out.

The Hunter Valley experience shows that the balance between resource development and farming/conservation has swung towards development in the extreme. Australia risks having many more Hunter Valleys, in places like the Darling Downs, the Galilee Basin and the Liverpool Plains. As the Warkworth PAC conceded, there is nothing in state law to protect NSW rural communities, and the laws of other states are little different.

There is a question for the nation as a whole here, too. As the world's population grows, making food production increasingly valuable, it seems contrary to Australia's national interest to be sacrificing prime farmland for short-term gain. In 50 years time, the growing populations of Asia may find cleaner ways to produce energy and in this case would no longer need our coal, but it's very likely that they will demand much more of our food production – if Australia is still able to produce it. Instead of seeing the strategic value of farming, politicians are blinded by the revenue hit they get every time they approve a new mine or gas project or expand an existing one. What binds together these two rushes – the CSG wells and the mega-mines – is the complicity of governments in maintaining weak regulatory oversight and in giving the resource sector special treatment so as to feed a growing addiction to mining revenue.

PART 2: POWER, INFLUENCE AND RESULTS

SEE NO EVIL

Mining and energy companies operating in Australia say that they are the most tightly regulated in the world, and they are right – in theory. Governments don't rubber-stamp new projects: they attach rafts of conditions that set limits for an array of environmental impacts, including toxic emissions, dust, water discharge and noise. As governments have begun approving mega-mines and risky CSG developments, the number of conditions attached to each project has risen inexorably.

Extensive lists of conditions make for impressive-sounding press releases issued by ministers when they approve projects. Interestingly, though, the mining companies don't complain about the conditions, largely because they know that governments rarely enforce them. The reality of Australia's so-called world's best-practice regulation is that both state and federal governments lack the willingness *and* the capacity to enforce the environmental limits set out in their approval criteria. This allows companies routinely to exceed limits without incurring a warning, let alone a fine.

As mining projects have become bigger, more complex and more unpredictable, ministers have even begun approving them subject to management plans not yet developed.

This was the case with the state and federal approval of BHP's Olympic Dam open-cut uranium mine, which is destined to become the world's biggest man-made crater. Contrary to rules applied to other uranium mines, BHP won approval for an open-cut expansion that will produce an above-ground 44–square kilometre storage facility for radioactive waste. This waste will remain active for 10,000 years and the entire operation will leach up to 8 million litres of contaminated waste water into groundwater each day for the first decade of operations, before falling to three million litres a day. Ken Buzzacott, an Aboriginal elder of the Arabunna Nation, took the federal government to court to challenge such an open-ended approval, but in April 2012, Justice Anthony Besanko of the Federal Court ruled against him, strengthening the right of the government to approve mines in this way.[73]

Even though the mining projects now being built or on the drawing board are worth $500 billion and are likely to exceed all the investment we have seen so far in this boom, major political parties are proposing not to strengthen regulation and monitoring, but to lessen it. The Opposition leader, Tony Abbott, has proposed a one-stop-shop policy that will hand over all of the environmental approvals to the states, subject to agreements with the federal government. Under new bilateral arrangements to be finalised with the states in 2013, Julia Gillard's Labor government has agreed with state governments to do something very similar, while reserving a role for itself in just a few areas which have nuclear or defence implications or involve Commonwealth waters.[74] These streamlined assessment procedures apply to developments in all sectors of the economy, although they have special significance for resource projects as these tend to have many environmental implications.

While there is no argument about the benefit of removing unnecessary duplication and excessive focus on the minutia as opposed to the real impact, there doesn't seem to be a strong case for reducing regulatory oversight of resource projects. It is not as though such oversight has inhibited development, given the flood of investment into the country. Yet the change is now occurring, even though the current regime – especially the role of the states – already does very little to inspire public confidence. Typically, companies set out their emissions levels in annual reports that are filed away in government departments, even when they show consistent breaches of limits. In far too many instances, companies are able to insist that the reports remain subject to commercial confidentiality so that they are never released to the public. In numerous instances people and communities living near mines have been denied access to monitoring results. Monash University's Dr Gavin Mudd, who has investigated many mining projects and their compliance, says that the 'vast majority' give 'commercial-in-confidence' as an excuse not to release compliance data.

The way that resources companies generate the information that goes into their reports is also questionable. In much of their reporting, they are able to choose the most favourable conditions to record levels of emissions, dust and noise. At times, they engage so-called independent consultants, but these consultants know that if they produce damning reports, they will not get hired again.

Occasionally companies do get fined: Whitehaven Coal was hit with a $6,000 penalty in March 2012 for polluting waters and breaching environment protection licences at its Narrabri and Tarrawonga coal mines in northern New South Wales. The NSW EPA treated these as serious breaches: "The EPA is very concerned by the number of incidents that have occurred and has formally put Whitehaven Coal on notice to improve its environmental performance."[75] But the $6,000 fine wouldn't even buy a grease and oil change on one big yellow truck. A few weeks later, Whitehaven was again fined by the EPA for yet more releases of waste water at its Moolarben mine: pollution that contaminated the Goulbourn River, which flows into the Murray–Darling basin. The EPA said Whitehaven had failed to put in place appropriate erosion and sediment controls to prevent runoff entering nearby waterways.[76] For this breach, Whitehaven was fined $105,000 – about the same price as a new tyre on one of those big yellow trucks.[77]

This easygoing regulatory regime has made Australia the darling of the global resource multinationals. With a combination of low levels of taxation, a compliant government and loose regulation, it's no wonder that the companies have plans to sink hundreds of billions of dollars of new investment into exploiting our national resources. The US giant Peabody Coal describes Australia as 'a premier location for coal mine development and investment'.[78] Likewise, the global mining services firm Behre Dolbear ranked Australia at the top of a study of 25 mining nations as a place to invest. In its 2012 study, the firm, which has provided technical and financial services to the mining industry for a century, said: 'There's no place better than Australia to put your investment dollars at [sic] work.'[79]

In a race to the bottom, mining regulation in Australia is a case of one rule for the miners and gas companies, and another for everyone else. State governments have, over several decades, competed with one another to win lucrative projects. As the mining boom has grown larger, they have become ever more competitive and willing to grant special treatment to the resources industry. State governments offer big projects exceptional status, such as being named a project of 'state significance' or being given approval under a fast-track route that avoids standard approvals. In New South Wales, for example, mines approved under the *Environmental Planning and Assessment Act 1979* are expressly exempted from complying with 17 state laws, including the *Native Vegetation Act*, the *Water Management Act* and the *Heritage Act*. This means miners can ignore regulations that prevent other land users, such as farmers, from cutting down trees, or that oblige them to curb their water use.

The problem with such fast-track approvals is that they often apply to several projects that impact on one another, thereby creating the risk of cumulative effects that go well beyond the risks outlined in a single Environmental Impact Statement. Most state laws don't require companies to take cumulative impacts into consideration.

To make matters worse, state governments have generally downgraded their independent environmental authorities, typically known as environment protection agencies, by rolling them into government departments answerable to a minister, and even by appointing boards stacked with industry representatives. This means that in most states, especially the resource-rich ones, there's little or no independent assessment of the mega-projects.

Working in these departments on approvals can be very tough indeed. Staff work around the clock when they are given tight deadlines to assess projects; they are often told, 'The minister wants this one approved.' The pressure from companies to get approvals through is often intense. Executives typically argue that their project's viability is marginal or that it relies on a marketing 'window' of opportunity that is about to close: without fast and lenient approval, the project will fall over. Senior public servants, especially in Queensland, say they have been put in humanly impossible situations in working on CSG approvals.

While the new LNP government in Queensland has placed some limits on the expansion of the Abbot Point coal export terminal inside the Great Barrier Reef and stopped two mines on prime farmland, other signs are not encouraging. A series of emails relating to the state government's May 2012 approval of the GVK–Hancock Alpha coal mine in the Galilee Basin suggests that industry is still able to wield undue influence and use an army of experts to get decisions rushed through. According to leaked emails from the office of Queensland's coordinator-general, who approves major projects in the state, officials told their counterparts in the federal environment department on 25 May 2012 that the government was set to give conditional approval to the mine, even though work on assessing the project's full impact under federal law was still some way off:

> We still seem to be a long way short of what you are looking for. I expect that Hancock will be lobbying heavily to obtain their approval from you once our report is finalised; they have had a direct line to the new government and the Co-ordinator-General here. On Tuesday [22 May] they came in with 22 experts to 'discuss' the proposed conditions, 48 hours before the report was supposed to be finished.[80]

The state government's approval of this mine is the first of the new model that hands responsibility to the states. The environment minister, Tony Burke, described Queensland's assessment of this project as 'shambolic' and suspended federal approval, but there's no sign of a retreat from his policy of introducing a leaner and faster approval process for these mega-mines.[81] Burke says the 'environment gains nothing from duplication or delay', even though more cautious approvals allow time to consider the long-term environmental consequences.[82]

States are being given more responsibility for approvals, yet their financial commitment to resource sector regulation and monitoring is very modest indeed – and is not linked to the amount of activity in the sector. Western Australia's Department of Mines and Petroleum is a big operation – it had about 800 staff and a budget of $65 million in 2011 – but this funding works out at just 1.7 per cent of the $4.9 billion in royalties collected by the state government from the resources industry.[83] Add in the other taxes paid by mining and energy companies in the state, and the share is much lower. And $65 million, down from $71 million in the previous year, represents a miniscule 0.04 per cent of the $147.5 billion worth of new projects being built in the state at the time.

In New South Wales, the approval of major projects is in the hands of the Planning Assessment Commission, a statutory body that received just $1.6 million in funding in 2010–11 to assess 43 major projects. That works out at about 1 per cent of the $1.4 billion in royalties and taxes paid by resources companies in that year, or $372,000 for each assessment of major projects that have far-reaching consequences for the citizens of the state.[84] Interestingly, PAC's 2010–11 annual report says it didn't engage any expert consultants in that financial year.

State departments are being raided by resources companies for their best and brightest, thereby undermining the capacity of the government to regulate this burgeoning industry adequately. Although Queensland boasts of its tough regulations, its environment department lost 70 staff to resources companies between January 2010 and early 2012, as its director-general, Jim Reeves, confirmed.[85] Formerly known as the Department of Environment and Resource Management (DERM), the department has since been renamed the Department of Environment and Heritage Protection (DEHP) by the new government: a name that puts even less emphasis on regulation and monitoring.

In January 2012, the Adelaide-based Santos, which is building one of the three big CSG projects, raided the department for some of its most experienced petroleum experts. The company poached Andrew Brier, at the time the head of a newly formed group designed to monitor the CSG industry, the LNG Enforcement Unit. Brier told friends that he was offered an astronomical package. He was joined at Santos by a unit director with the LNG Enforcement Unit, Jim Belford. The third Santos recruit was Rod Kent, who headed the petroleum and gas, environment and natural resource regulation division within the Department of Environment. If this was a case of public servants applying for advertised positions, then there would be nothing amiss, but Jim Reeves confirmed that resources companies have been actively courting his experts with job offers.

And with what result? When Dean Ellwood, the department's assistant director-general for natural resource regulation, was asked how many production wells have been inspected by the new LNG Enforcement Unit, he highlighted the department's monitoring of 320 water bores that had shown no ill-effects from CSG activities. (A few months later, the Queensland Water Commission [QWC] said that more than 600 bores would indeed lose water due to CSG operations.) When asked what regulations had to be followed to frack wells, Ellwood indicated that companies simply had to fill out some 'risk assessment' forms. His response underscores the DIY approach to regulation that most state governments rely on in Australia. He explained:

DERM requires businesses seeking approval to undertake fraccing [sic] activities to submit a hydraulic fracturing risk assessment that considers more than twenty-six potential risk factors. These include details of the proposed chemicals to be used, the toxicity of those chemicals and their mixtures, practices and procedures to ensure the fracc [sic] is contained within the target area, and environmental and human health hazard and impact assessments. Hydraulic fracturing would not be authorised where this risk assessment has not been carried out to DERM's satisfaction.[86]

Ellwood's answer implies that DERM didn't do a great deal of on-the-ground monitoring, relying instead on what is known in the trade as desktop analysis – or sitting at a computer. When asked again to outline specifically how much monitoring was done, the department boasted of 'a significant field presence' that in 2011 involved 99 CSG field visits. On each visit, inspectors looked at 'up to 15 wells', along with other infrastructure, such as dams, compressor stations and reverse osmosis plants.[87] Assuming that they did the maximum of 15 wells on each visit, that's still less than 1,500 wells a year, or one-third of the existing number of wells in a state that anticipates exponential growth over coming decades.

If CSG were an industry such as pharmaceuticals, government would require companies to conduct small-scale lab-based tests to assess the potential risks and side-effects of new ventures. But when it comes to mining and energy, no such precautions apply. Drug disasters like Thalidomide were quick in their impact, and governments responded, but the potentially damaging effects of CSG projects on groundwater may take decades to show up. In the meantime, state governments will be cashing in on their share of the production revenue via royalty agreements. And, at present, they can't authorise new projects fast enough.

JUST CAN'T SAY NO

In the early 1970s, Australia became the second country in the world after the United States to require mining companies to complete environmental studies as a pre-condition for project approval.[88] The federal government introduced legislation modelled on the *US National Environment Policy Act* of 1969, and then state governments introduced their own legislation.[89] Despite the battery of procedures, as few as 1 or 2 per cent of projects at any time ever get knocked back, even though the scale and potential impact of such projects has grown immensely over the last 10 years.

The CSG developments approved by the Queensland and federal Labor governments are a prime example of the practice of attaching large numbers of largely meaningless conditions to risky projects in order to sell their approval. Less than two months after federal Labor regained office in 2010, the newly minted environment minister, Tony Burke, granted 50-year licences to two enormous CSG projects that involve $35 billion of investment, almost 10,000 production wells and just as many thousands of kilometres of pipelines and roads, all linked to two new gas processing plants built within the Great Barrier Reef Marine Park. A few months later, Burke approved a third project equal in size to the first two combined. Burke's Department of Sustainability, Environment, Water, Population and Communities (from here on abbreviated to the federal Department of Environment) advised him that after approving the first two projects, he just couldn't say no to the third, ignoring the cumulative impact. These were three extraordinary decisions: conservationists and farmers alike argue that these mammoth projects pose the greatest risk to prime farmland, water, food production and farming communities that Australia has ever seen.

Tony Burke was a committed environmentalist in his early years, when he joined the Wilderness Society while studying law and campaigned to save the Daintree Rainforest in Far North Queensland. However, as a true son of Labor's NSW Right, Burke has since adopted what might be called an 'icon' strategy. This means protecting a clutch of high-profile environmental sites – such as marine parks, Cape York, the Kimberley and Tasmania's forests – in the hope that these achievements will overshadow the damage wrought by the large-scale developments he has approved. Burke has openly canvassed a strategy for protecting these high-profile sites with the green groups that stand to gain from it. This strategy sounded compelling to some greenies before the potential damage to the Great Barrier Reef from the Galilee Basin mega-mines became apparent. Burke's next big headache will be decisions on these three big mines that threaten the mother of all environmental icons, especially after UNESCO warned in its June 2012 report that these developments could put the reef onto its List of World Heritage in Danger.

One of the big CSG projects approved by Burke in October 2010 is Gladstone LNG (GLNG), a consortium led by Santos. As with Olympic Dam, many of the reports and management plans on how the project will address environmental hazards will only be provided *after* approval has been granted; some are still yet to be provided. When I asked the federal Department of Environment how GLNG had complied with some of the conditions imposed by Burke, it emerged that the project's compliance is conditional on a range of technical and bureaucratic processes that are all still ongoing. For example, when asked if Santos had provided a copy of cumulative impact report, it sounded as though Burke's department was caught in a bureaucratic merry-go-round, as it was unable to say when this report would be provided:

> The cumulative impact report is to be provided to the Federal Government when it is provided to the Queensland Government. The timeframe for Santos to provide this report to the Queensland Government is a matter for the Queensland Government.[90]

When asked about the project's groundwater assessment, as required by Burke, the federal Department of Environment said that this critical study, which has enormous implications for the nation's water resources, was being handled by Queensland's coordinator-general, and that 'The timeframe for Santos to provide this report is a Queensland Government matter.'[91] Apparently unbeknown to Burke's department, Santos did the study as required, but it has not been released because the state government decided unilaterally that this condition could now be fulfilled by the QWC study. 'The original Santos/GLNG modelling and assessments are no longer required. The proponent is not required to take any further action because it has already met the condition in the CG's report,' said a spokesman for the Queensland department of state development.

Regarding the management of the water generated by gas production, Burke's department was unable to say when the company would release its 'CSG Water Monitoring and Management Plan' for stage one of the project, yet another document apparently required by Burke when he approved the project. The Department of Environment said that it was provided with the plan in April 2011, but almost a year later this plan was still being reviewed by an 'expert panel'. Once it is approved by the panel, Santos is required to publish the report.[92]

When Burke's predecessor, Peter Garrett, became environment minister in 2007, the former rock star brought to this role a committed background of three decades of musical activism in support of the environment and Indigenous rights. But as a Labor minister, Garrett quickly showed a capacity for political pragmatism in dealing with the multinational giants he had formerly railed against. Just over a year after becoming minister, Garrett faced the decision of whether or not to approve the expansion of a lead and zinc mine at McArthur River, arguably one of the most contentious and potentially damaging projects of the past two decades. The process that eventually brought this decision to Garrett is an example of worst-case shoddiness.

The mine is located 65 kilometres from the largely Aboriginal town of Borroloola, 10 hours' drive south of Darwin in the Gulf of Carpentaria. Its expansion involved diverting the McArthur River a distance of six kilometres into a man-made channel to allow the Swiss-based multinational Xstrata to mine the riverbed. Previously the mine had operated underground with minimal impact but – in an all-too-familiar trend – the boom in prices made it compelling for Xstrata to propose an expansion.

The first assessment by the NT EPA was rigorous and resulted in a rejection of the proposal by the NT government in February 2006. The EPA expressed serious concerns about the impact and warned of eventual destruction of the man-made channel. It said the project could have an adverse impact on the river and that it was concerned about long-term management of sediment and contaminants.[93] It warned that Xstrata's approach to environmental management meant 'shifting the risk' from the company to the government. But a month later, the government asked Xstrata to re-submit its application after the then chief minister, Clare Martin, declared she was 'keen to see the mine operate'. In the second review, a more rigorous process did not apply, says Kirsty Howey, a solicitor who worked on this dispute for the Northern Land Council. In October that year, the NT government approved the mine and a week later the federal environment minister, Ian Campbell, did the same under federal legislation. As Howey concluded in a detailed account: 'The flexibility of the Northern Territory and Commonwealth's environmental assessment processes was used to the maximum effect to secure approvals in the face of a wealth of scientific evidence on the adverse impacts of the proposal.'

Xstrata promptly began work on shifting the river, but in April 2007 the Northern Land Council, acting on the wishes of the local Aboriginal people, won an action in the federal court to halt the open-cut expansion and river diversion. The court ruled that the process leading to the approval was flawed. But this victory was short-lived. The NT government responded the very next day by changing the law to allow the river diversion to go ahead.

In the aftermath of the decision, Peter Garrett flew to Borroloola ostensibly to consult with the local elders before making a parallel decision on whether the mine should proceed under federal legislation. It was a big day when the bald-headed rock legend arrived in the town; no-one could remember the last time a federal minister had visited Borroloola, and here was one who had previously sung in support of Indigenous rights.

Federal environmental laws require review of the NT government's authorisation under the *Environmental Protection and Biodiversity Conservation Act 1999*, the main piece of federal law that's used to assess whether a project will harm natural heritage of 'national significance'. Although Garrett had the power under this Act to either halt or approve the open-cut expansion, he was already back-pedalling by the time he arrived in Borroloola. 'The hardest thing, I think, for everybody to come to terms with is that my role under the Act is relatively limited,' he said on the day.[94]

Garrett met with a group of elders at the NLC's Borroloola office. When he sat down in the courtyard under an ochre-red shade cloth, the locals knew their fate was sealed. The former stockman Jack Green, an Indigenous Australian who now works on a carbon abatement project, said everyone there knew Garrett was going through the motions. 'As soon as I seen him, I just knew he was going to approve it,' he said.[95]

Garrett explained that he couldn't put the river back:

> Putting the river back is not part of my decision-making process here. The decision-making process is about whether or not the mine, as it's constituted, as it's been proposed and, in fact, as it's been operating, has impacts on matters of national environment significance, which are identified in the legislation.[96]

Given the shoddiness of the approval by the NT and federal governments in 2006, this was a project screaming out to be halted. Garrett could have found a way to leverage his limited powers, but he balked at what would have been a bold and defensible move.

Exactly a week after the visit, Garrett announced that he would approve the mine's expansion 'subject to strengthened conditions'. The nine conditions are designed to minimise the impact on the nationally protected freshwater sawfish and ensure that shorebird populations near Port McArthur – where the lead and zinc concentrate is loaded for export – will be properly monitored.

Aboriginal elders in Borroloola seem saddened by what has happened. While they have tolerated the underground mine since 1995, they say the open-cut operation is affecting the McArthur River and sacred sites located near the mine. They attribute a spate of deaths of several elders in recent years to the river diversion. They believe that the mine must be discharging toxins into the river because fish have been dying in the water, while those caught have shown what the elders call 'black fat'.

The federal Department of Environment says there have been no instances of non-compliance with the mine's conditions, as confirmed by an April 2010 audit that identified a 'high level of compliance'. But when asked if the report could be released, the department said that 'audit reports are not generally made public' and the government would need to seek permission from the company before releasing it. The department said it had no further plans to conduct another audit of the mine's compliance. End of story.[97]

It is hard to conceive of a poorer process at both the state and federal level. Not only does a highly contentious mine get approved in the shoddiest way possible, but the public also has no access to the monitoring of its compliance with the conditions imposed, and nor is there a regular monitoring program.

CASH FOR COMMENT

The voluminous Environmental Impact Statements produced for resource projects often run to thousands of pages and create an aura of appropriate scrutiny, even though remarkably little independence goes into their creation. The resources companies develop their own EISs by commissioning reports from so-called expert consultants who have a vested interest in getting more work. These reports are rarely subjected to peer review, and if they are, the panel is usually a friendly one. The system has an inherent and dangerous bias. Not only do resources companies engage consultants who are likely to produce favourable reports, but these companies also control and vet the information that goes into the EISs. Consultants who produce negative reports won't get work again.

Some of the few environmental experts who aren't on this gravy train say the whole process is deeply flawed and in need of radical overhaul. Barry Carbon, a former CSIRO scientist and former head of Western Australia's EPA, says the EISs are often so long that bureaucrats don't have time to read them. In addition, there is no independent accreditation of the consultants who produce them. The NSW government recently rejected a long-overdue reform put by a Greens MP, Cate Faehrmann, to require the consultants to be accredited. The process is so open-ended, such a free-for-all, that companies can submit whatever level of detail they want. Some CSG projects have submitted as few as eight pages of analysis on their impact on groundwater, others over 1,000 pages.

The only check in the system is public scrutiny, but with the exponential growth of such projects, it is impossible for the media, environmental groups, independent academics and indeed government departments to read, digest and comment on all these reports.

Professor Mark Taylor of Macquarie University describes the system as 'cash for comment' – that is, akin to non-disclosed payments made to radio hosts for favourable endorsement. The problem is what he calls 'academic enterprise', whereby university staff are required to bring in dollars from whoever will offer them, irrespective of any conflict of interest or pressure to produce favourable reports. Often academics will refuse to write reports, be expert witnesses, provide evidence or make public comment for fear of 'losing their funding because they have made adverse comment or produce evidence which challenges the status quo'.[98] 'Academics are in certain cases no better than people in industry; they are part of the process of obfuscating information and the truth – either passively or deliberately,' Taylor laments. He says the consulting firms 'know that if they give an adverse report, they will never get the job again'.

Monash University's Dr Gavin Mudd argues that the environmental consulting industry lacks any form of independent certification and, like Taylor, says the reports should be produced independently. He cites the case of AGL's EIS for the Gloucester CSG project, which he describes as 'a shocker'. It had a 'pathetically minimal and weak' eight-page discussion of the potential impact on groundwater, and it omitted to include a 'cross-section' drawing that showed the interaction between the project and the aquifers and coal seams below the surface. 'If I were a consultant, I would be ashamed to put my name to it,' he says.[99]

The NSW Environmental Defenders Office (EDO), a legal aid service for environmental issues, singles out Santos's EIS for its CSG projects in New South Wales as an example of how the process is flawed. The Glasserton pilot slated for the rich farmland of the Liverpool Plains involves drilling three exploration wells. The EIS states that the water will be extracted from the Bluevale subcatchment, but in fact the project will affect a totally different catchment.[100] 'This is a serious error given the sensitivity of the aquifers in the area to drilling activities,' says the EDO. Nor does Santos mention, in its 'Review of Environmental Factors', the close proximity of the pilot wells to Goran Lake, a wetland that supports a variety of rare, endangered and vulnerable species.

Challenging these development-driven EISs is an expensive and time-consuming business for affected communities. Getting access to information in the first place is often not easy. According to Queensland's EDO, the state's residents have more right to information about a simple home extension than they do about a mega-mine.[101] Under recent changes that apply to CSG and mining generally, state government websites list opportunities for submission or objection, but there's still no complete set of documents available online. The Brisbane City Council requires property developers to put their application and supporting information on a central government website, but there are no such requirements with major resources projects.

Farmers and community groups who decide to put an alternative view can spend tens of thousands of dollars on independent reports, along with countless hours of their time. The Liverpool Plains farmer Tim Duddy has spent five years working full-time to halt the spread of mining to his farming region. Residents in mining-affected areas of the Hunter Valley, who have been challenging the relentless expansion for decades, have entire bedrooms devoted to filing the reports issued by companies.

The race for riches is putting ever greater pressure on consultants to sign off on impact statements that are travesties of omission and data manipulation. Put simply, the easiest thing for consultants to do is not to find endangered species when they conduct a study of a proposed mine site. They can visit the site during the day, for example, instead of at night when the animals come out. Or when consultants are asked to review a mine's impact, they can put the disappearance of swamps and lakes down not to the longwall mining underneath, but to low rainfall.

The extent to which environmental consultants are willing to give resources companies exactly what they want is revealed in stark terms by the experience of the ecologist Ray Mjadwesch, who could have made himself a very rich young man in the 12 years that he worked in the ecological consulting industry. Shortly after graduating from university with majors in biology and ecology, Mjadwesch got his foot in the door with a study in the western coal fields of New South Wales.

Longwall mining involves extracting horizontal panels underground that follow the coal seams, between three to seven metres thick, 250 to 450 metres wide, and up to 10 kilo-metres long. Clive Palmer's China First Coal Project is one such mine. The process has a number of serious risks, most notably land subsidence. Such mining has been blamed on the loss of Thirlmere Lake in New South Wales, which used to sit above a longwall mine operated by Xstrata.

Reporting on the impact of his company's longwall mining, Mjadwesch discovered subsidence: there were cracks wide enough for a cow to fall down, and a swamp had dried up. He raised these findings with the company's environmental officer, who told him: 'No-one is telling us this – you are opening Pandora's box.' When Mjadwesch put the findings in his report, the environmental officer convinced him to change a few key words. Instead of writing 'severely subsidence affected', for example, he wrote 'subsidence affected', as the severity of impact had not been quantified.

This job led to a really big opportunity, a contract that was to run for five years beyond the life of the mine, to review the impact of another longwall mine nearby. This mine had been operating for almost a decade and the company in charge wanted it to continue, but the state government reminded the company that it had not yet furnished the monitoring reports on the impact of its operations, which had been promised when it was approved in 1992.

When Mjadwesch first looked at the area affected by the mine, he noticed that swamps located above the operations had dried up; he put this finding into his first report, which he delivered six months after starting the job. Shortly after he filed the report, Mjadwesch was called to a meeting by the company at which a fresh group of consultants was present. One of the consultants told Mjadwesch that he had worked on 50 EISs, all of the projects had been approved, and that no-one had ever predicted or reported such impacts. When Mjadwesch showed the consultant a drained swamp, the consultant argued that the water loss could have been caused by lower rainfall. Mjadwesch then analysed the rainfall data and found there was no connection, so he stuck to his findings.

Like many consultants, he was obliged to report to the state's National Parks and Wildlife Service (NPWS). Unlike many consultants, who cite 'commercial in confidence' as their excuse, he carried out this task. He had professional obligations as a registered ecologist with NPWS, and as a result felt compelled to submit his report. After he made the submission, NPWS called to tell him that someone had come to the office to retrieve the report, saying that it was a 'draft report which a consultant had submitted in error'. A few hours later, he received a fax from a law firm terminating his involvement in the monitoring project. Since then, the mine's consultants have consistently reported 'no impact of subsidence'.

Over the past decade, Mjadwesch has continued to campaign for action to protect the swamps; he has written numerous submissions to the government on the impact of longwall mining on this ecologically rich environment. Only in response to irrefutable evidence of impacts, including inspections by botanists and field trips by the NPWS Biodiversity Conservation Unit, did the NSW government finally act by slapping an Enforceable Undertaking on the mine: a proxy fine of $1.45 million. The company responded by pitching this as its 'contribution to research', promising to do even more monitoring, and narrowing the width of its longwall panels.

At the heart of the problem experienced by Mjadwesch is the absence of any independent certification of the experts who produce the reports. 'The problem is that the EIS consultants don't have to be registered in the same way that people who produce estimates on ore bodies have to be certified by JORC [the Joint Ore Reserves Committee],' says Gavin Mudd. One option is to certify consultants, but a far-reaching reform would see governments take responsibility for managing and certifying the quality of these reports. 'Governments have to take responsibility – charge companies for doing them, or at least certify the system for producing EISs. The EIS needs to be an independent process,' Mudd says.

Over and above the approval process, an even bigger problem is the absence of independent and transparent monitoring. DIY regulation involves submitting reports to state government that often include breaches of conditions, with no consequences. When government agencies do act, as we have seen, the penalty is something like a $6,000 fine. The environmental scientists Mandy Elliott and Ian Thomas argue that while monitoring is part of the environmental assessment process, it 'is seldom emphasised'. They argue that follow-up mechanisms remain poorly developed and external review of assessments is 'negligible': 'Even where monitoring is specified as one of the matters to be dealt with, surveillance of the monitoring programs is rare.'[102]

The solution to these problems has to be stronger and more independent institutions that can attract and retain experienced technical staff by paying competitive salaries. Without experienced public servants, resources companies will put government and communities at an extreme disadvantage in protecting the public interest. Such a reform would involve developing much larger environmental regulators, possibly in the form of independent statutory authorities. While this might sound like a difficult reform to put in place, a similar system is already operating in Australia for offshore oil and gas projects. Ominously, it took Australia's worst oil spill in a generation to spur the federal government into action to ramp up its regulation of this industry. Up until the August 2009 Montara oil spill in the Timor Sea, regulation of offshore rigs had been in the hands of the state and territory governments. The inquiry into the disaster that created a 6,000–square kilometre oil slick found that the NT administration had relied on self-reporting by the companies without any independent verification of their claims about adherence to regulation – very much like the model that exists for state-based regulation of mining and CSG.

In the wake of the Montara spill, the resources minister, Martin Ferguson, put stop-gap measures in place by giving responsibility for the regulation of oil wells to the National Offshore Petroleum Safety Authority, a body that had previously been responsible for the health and safety of workers. Then, in January 2012, a new national regulator came into being, the National Offshore Petroleum Safety and Environmental Management Authority (NOPSEMA), which has oversight of occupational health and safety, wells and well operations, the structural integrity of facilities, and environmental management within Commonwealth waters.

Despite the dogged refusal of Western Australia to support a national regulator, Ferguson was able to use the Montara disaster to achieve far-reaching regulatory reform. Significantly, NOPSEMA is a statutory authority with its own independent advisory board. It operates on a cost-recovery model that means charging industry for its services, and it pays 'highly competitive' salaries to its technical experts, thereby preventing loss of talent to lucrative jobs in the industry. For example, a well-integrity specialist – a job that involves certifying that wells have been properly drilled – can earn a salary of $231,000 a year. NOPSEMA shows that Australia has the capacity to create strong statutory institutions.

The sheer scale, complexity and speed of resource development in Australia calls for more vigorous and effective government oversight, and much greater public disclosure by resource companies. Given the mega-mines and CSG projects now being rolled out, there's a serious risk that the current set-up will allow more Montaras to occur – but with much greater and longer-lasting ramifications. The case for robust regulation is compelling, but given the power and influence of the resources industry, there seems to be a dearth of political will to give Australians the type of approval processes and oversight they deserve.

THE TAIL THAT WAGS THE DOG

John Anderson is a former deputy prime minister of Australia who comes from a farming family that has worked the rich farmland around Gunnedah in northern New South Wales since the mid-1800s. When he entered federal parliament in 1989 at age 33, this good-looking, well-spoken and upstanding Christian seemed destined for great things. He went on to serve as minister for transport in the Howard government and then, after a decade in politics, succeeded Tim Fischer as leader of the National Party – the party of the bush, once known as the Country Party. Anderson served as deputy prime minister for six years, before suddenly resigning in June 2005. He was replaced by another National from northern New South Wales, Mark Vaile, who had worked as a real estate and stock agent before entering politics. Youthful and professional in their appearance and outlook, Anderson and Vaile presented a fresh image for the Nationals.

The National Party is meant to be the political force dedicated to serving the interests of people on the land, but the career decisions made by Anderson, Vaile and several other 'Nats' immediately after they left politics underscore the shifting power balance in Australian politics and the rising influence of mining. Instead of continuing to work for rural businesses and interest groups, Anderson and Vaile leapt the paddock fence and went to the other side, standing up as front men for the miners and gas producers at a time of rapid expansion.

The Nats in government are now showing a surprising enthusiasm for mining, even when it comes to the best farmland in the country. Two months after winning office in 2011, the NSW Nationals leader, Andrew Stoner, asked for a letter to be drafted to the planning minister, Brad Hazzard, regarding the Shenhua coal mine slated for the Chinese state company's 195–square kilometre lease on the Liverpool Plains. He asked that the letter highlight the 'jobs, investment and infrastructure' that will result from the project, according to an email from his policy director, Alex Bruce. Incredibly, Stoner asked that environmental approval not be held up by the Namoi Water Study examining the impact of mining and CSG on groundwater resources. Stoner also instructed that the letter should state his belief that the project 'is of State significant [sic] and is a high priority and in light of this should be progressed along the assessment path as quickly as possibly'.[103] Such breathless enthusiasm for a 300 million tonne open-cut mine on the best farmland in the state could be expected from a mining company executive, but coming from the Nationals leader it was truly staggering.

In 2007, John Anderson joined the board of the CSG company Eastern Star Gas (ESG) as non-executive chairman. ESG was an emerging CSG producer based in Toowoomba that has come to symbolise how poor oversight can prevail under a DIY approach to project monitoring. While Anderson was chairman, ESG was responsible for a succession of spills of saline and possibly contaminated water from its wells in the Pilliga Nature Reserve, located just 20 kilometres from Anderson's Mullaley property. One spill in June 2010 involved 10,000 litres of water that affected 1.2 hectares of bushland. The results of the spill are devastating and can still be seen more than two years after the event. Then, in January 2012, a further 250 litres of the algae suppressant Algaecide was spilled at ESG's Bibblewindi water treatment plant.

The charges concerning ESG's responsibility for these lapses were not only made by conservationists, who have posted footage of the filmed devastated area on YouTube. In November 2011, ESG was acquired by Santos. The new owner released a detailed report in an effort to set the record straight. A soil analysis found that a small area of vegetation adjacent to the Bibblewindi plant has been poisoned by the elevated salt levels. This was the result of accidental water releases at the site 'which should not have occurred and would have been preventable if Eastern Star had appropriate processes in place'.

Santos committed to spending $20 million to upgrade the operations, but the company has retained many of the workers who were on board at the time of the spills. Anderson's public accountability for this incident has so far been nil. Following the release of Santos's damning report, which was sent to the NSW government, he declined to comment to the media.

Mark Vaile's foray into the resources sector has involved adding his name to the board of Aston Resources, a company majority-owned by the former electrician turned billionaire

Nathan Tinkler. Aston is behind plans for a coal mine at Maules Creek, which it acquired at a firesale price from Rio Tinto. It has plans for an open-cut operation with a capacity of 13 million tonnes per annum that will carve out half of the Leard State Forest, a wildlife sanctuary amid the rich farmland of northern New South Wales. The company's EIS says the forest is home to 22 mammal, 132 bird, eight frog and 25 reptile species. The other half of the forest will be taken out by the expansion of the adjoining Boggabri coal mine, owned by Japan's Idemitsu Australia. Boggabri is typical of so many mines in Australia; it began life as a 1.5 million tonne operation, but now has plans to go to 7 million tonnes a year.

Aston has already secured farmland on the site of the proposed mine, even though it is yet to be approved by the state government. After buying Maules Creek in 2009 for $480 million, Aston has since substantially increased its estimate of the reserves to more than 600 million tonnes of proven coal, with marketable reserves of 320 million tonnes. If coal prices remain above $100 a tonne, the mine could earn revenue of $1.3 billion a year; and if Mark Vaile's remuneration is only a tiny fraction of that amount, he will have done very well indeed.

The Maules Creek development has put Tinkler and Vaile in line for greater things. In mid-2012, Aston merged with Whitehaven Coal to create a $5 billion company, with Vaile as chairman and Tinkler as a 20 per cent shareholder. Whitehaven owns most of the coal mines and prospects around the Gunnedah region in northern New South Wales, but the company's approach to environmental management has also attracted the attention of the NSW EPA for a series of breaches.

Farmers in the region affected by the Maules Creek development are angry with the way the former Nationals leader has sided with the miners. Phil Laird, whose family has farmed in the area since the mid-1800s, expressed bitter disappointment with politicians as he spoke on the edge of the Boggabri mine while a bulldozer was tearing down the ancient forest:

> Our family has been Coalition voters since the day dot. My father, who is near 80, mans the local Maules Creek polling booth for the National Party handing out How to Vote material for state and federal elections and supporting people such as John Anderson. The sense of betrayal of country people is palpable. When we found out that Mark Vaile was the chairman of Aston Resources, we couldn't believe it. It is fair to say that any faith that we have had in our politicians to represent our interests has been seriously shaken.
>
> Our family has been in the district for 150 years and we are conservative people who rely on our farms, surrounding landscape and climate. It is clear to us at least that clearing a forest in order to develop a coal mine is completely wrong and out of step with current thinking. In a spirit of compromise we have met with Mark Vaile and have urged him to build an underground mine to protect our environment and our community. However, these requests have fallen on deaf ears.[104]

Vaile contends that his involvement with mining reflects his concern about the 'hollowing out' of rural Australia over recent decades. He believes that mining will be able to reverse this trend and provide greater opportunities to people. He says that the conflict between mining and farming does not have to be 'mutually exclusive, it has to be balanced', although he admits that the Maules Creek mine will cause some farmers to 'leave the area'.[105] Vaile argues that Maules Creek does not involve building a mine on prime farmland like the Liverpool Plains. He describes the land in the area as 'third class', with many rocky ridges, and says that 'some of it is poor country'.

Laird responded to these comments by pointing out that the project will destroy something he believes is far more valuable than prime farmland that has been cleared for intensive agriculture. 'This is not black soil cropping land devoid of life but is a large intact remnant of the White Box grassy woodland sitting on the edge of the extensively cleared Liverpool Plains and is unique in this context,' he says.

The involvement of the Laird family in protecting the forest shows the depth of commitment required if individuals and communities want to resist mines. Laird, a lanky IT consultant who has worked extensively overseas, returned home to the farm about a decade ago with his wife to raise their two boys, but has since been embroiled in the fight against the two mines and the prospect of CSG operations. The region is covered in petroleum exploration leases (which include CSG activities) that extend from Sydney all the way into the gas fields of Queensland. The leases around Gunnedah are owned by AGL, Dart Energy and Santos.

Laird helped to set up the Maules Creek Community Council, which has spent more than $40,000 in getting independent assessments to challenge the miners' claims in their EISs. It seems an uphill battle. The state government has set up a process that will make it very difficult for the people of Maules Creek to appeal to the Land and Environment Court. Laird also realises that he is up against a formidable lobbying effort led by Vaile. He thinks that Rio Tinto sold to Aston because it believed state government approval would be difficult, but Aston's political connections could well make the difference. The NSW Greens' mining spokesman, Jeremy Buckingham, says that Vaile is seen around Parliament House in Macquarie Street 'again and again' as he pushes for approval.

A big part of what's involved in getting new projects over the line seems to be extensive donations to political parties and to the local community. In 2011, Nathan Tinkler, now a resident of Singapore, gave $50,000 to the federal and NSW Nationals despite having ownership of a property development company that would make such donations unlawful. The donation is being investigated by the NSW Planning Department, which stated that Aston had twice failed to declare its donations: in 2010 when it submitted its development application; and in November 2011 when the Planning Department contacted the company to confirm that there were no donations to declare.

Tinkler's failure to declare his company's donations might reflect his attitude towards public scrutiny. When a *Sunday Age* journalist, Tom Reilly, telephoned Tinkler in 2010 to ask about his acquisition of half of Dick Johnson's racing team, Tinkler told him: 'You're a fucking deadbeat, people like me don't bother with fucking you. You climb out of your bed every morning for your pathetic hundred grand a year, good luck.'[106]

Aston has also drawn on Vaile's political expertise and shown that it understands how to work the local community with carefully targeted donations. Despite not having government approval for the project, Aston has put air-conditioning into the local school and new equipment into the Boggabri hospital, while donating $20,000 to the Maules Creek campdraft, a sporting event that involves horseriding and cattle herding. Vaile says his 'political background' influenced the company to engage with the community 'while going through the environmental approval process'. The company did a 'lot of preparatory work on the ground' that involved 'deeply engaging with the community'.

Rose Druce, a third-generation farmer, sees things differently. She says Aston has been trying to silence the community with its donations: 'To me, that's shut-up money.'[107] Asked if Aston was trying to buy the community, Vaile conceded that some people would make that charge, but he said: 'We are not, I have just always taken the view that … the company is going to be part of the community for many, many years. There has got to be benefits back to the community.'

Anderson and Vaile are joined by a phalanx of former politicians and political advisers in the employ of the miners. Like no other industry, resources is recruiting a vast network of former politicians and hangers-on to get projects over the line, and by and large the companies are succeeding. Given the intersection of mining interests and agriculture, the National Party has come in for special attention. But the linkages with the Liberal Party are more extensive and also appear to be delivering results.

Stephen Galilee, a former adviser to federal Opposition Leader Tony Abbott and the state treasurer, Mike Baird, is chief executive of the Minerals Council of New South Wales. Galilee is a Coalition adviser of long standing, having worked for the party during opposition days in the early 1990s. Should Abbott succeed in becoming prime minister, Galilee's stocks can be expected to rise even further. The decision by the NSW government in February 2012 to back away from its election commitment to protect prime agricultural land from mining and CSG developments seems to be a textbook case of how the industry is winning the lobbying game. After the release of the draft regional land-use policy, shares in Santos and AGL – the CSG companies most exposed in New South Wales – staged a rally, indicating that the market was relieved about the policy.

The fledgling CSG company Metgasco also has good political connections. Its corporate affairs director is Richard Shields, who worked as an adviser to the federal communications minister Helen Coonan from 2003 to 2006, before moving to the NSW head office to work in a fundraising capacity, where he later became deputy director. Metgasco has three production exploration licences that cover a large part of New South Wales's lush northeast, taking in the towns of Casino and Grafton. Another Liberal recruit, the former

Howard media adviser Ben Mitchell, is communications director at the Minerals Council of Australia.

This network strongly influences Coalition policy and statements, including the words that come out of Tony Abbott's mouth. During the dispute over the super-profits tax in 2010, Abbott declared his undying loyalty to the industry when he vowed to fight the proposal while there was still 'breath in my political body'. He has defended CSG by arguing that farmers benefit by getting better roads. And he has even advocated in favour of the exponential growth in coal shipments through the Great Barrier Reef and declared that yet another mega coal mine is vitally important for our future prosperity. When Tony Burke froze federal approval of the Alpha coal mine in June 2012, Abbott said:

> We've been exporting now through the reef for many, many years. Obviously you've got to make sure that there are appropriate safeguards taken, but look I think it's very important that we do expand, appropriately, coal mining in Queensland. We can't let a green veto stop these projects which are vital for our future prosperity.[108]

Significantly, these statements contain specialist knowledge which suggests that Abbott is being scripted by the industry. As Opposition leader he has shown an extraordinary willingness to advocate on its behalf. He even adopted one of its tactics – running down agriculture – when he said that run-off from farms had damaged the Great Barrier Reef. It is an absurd argument – that because one industry has damaged the reef, this is justification for another to expand massively and risk doing more damage.

Labor is not entirely out of the loop. Prime Minister Julia Gillard's former chief of staff, Amanda Lampe, had a stint with Origin Energy before joining the Gillard team. When Queensland's deputy premier and treasurer Keith de Lacy stepped down, he went straight to the board of MacArthur Coal. The Labor-connected lobby group Hawker Britton also secured work via its subsidiary Barton Deakin to lobby the state government on behalf of Metgasco.

The industry's best assets inside the Gillard government are the resources minister, Martin Ferguson, and the former national secretary of the ALP, Gary Gray, who went on to work for Woodside as a lobbyist. Ferguson is a hardworking and intelligent minister, but it can seem as though he has become 'orestruck'. He talks of a second pipeline of investment that could eclipse what is currently underway. While Australia is a magnet for such investment, Ferguson has warned the government that capital is 'footloose' and any changes to taxation would scare off the global giants.[109] The industry adores him. When he backed Kevin Rudd's failed leadership bid, the Minerals Council publicly counselled Prime Minister Gillard against dumping Ferguson as resources minister. For his part, the WA-based Gray left Woodside to enter federal parliament in 2007, taking the seat of former leader Kim Beazley. From the backbench Gray played a role in stirring up caucus unrest over Kevin Rudd's super-profits tax. Now, as special minister of state, Gray is a member of Cabinet, one whom the industry can rely on to ensure that their views are heard and heeded.

The WA networks seem even more entwined than those of Queensland and New South Wales, given the dominance of the resources sector. Premier Colin Barnett's former chief of staff, Deidre Willmott, has a senior executive role at Andrew Forrest's Fortescue Metals Group (FMG). It was Willmott who gave up her pre-selection for a safe seat when Barnett decided to return as opposition leader in 2008.

Barnett's Indigenous affairs minister, Peter Collier, has declared that Forrest is someone from whose 'wisdom' he has taken 'guidance and great advice'.[110] It was Collier who, in late 2011, agreed to FMG's extraordinary request to delete Aboriginal heritage conditions on its mining lease that covers an area in the Pilbara claimed under native title law by the Yindjibarndi people. The conditions deleted by Collier dealt with the need to avoid all sites containing Aboriginal human remains, to consult local Aboriginal corporations to identify heritage sites, and to provide the registrar with data on the location of all rockshelters and caves, including copies of all archaeological and anthropological reports. Collier's department also seems to have ignored FMG's refusal to allow the Yindjibarndi people access to their sacred sites, even though this is one of the conditions of the lease.

Mining companies are reinforcing their political networks with hard cash; the sector has emerged as the single biggest donor to political parties and continues to spend large sums on political campaigns. After spending $22 million to demolish the super-profits tax in 2010, cashed-up mining companies spent more than $8 million on political activities in the 2010–11 financial year, according to the Australian

Electoral Commission, with most of the largesse directed to the Liberal and National parties. The Minerals Council of Australia and the Association of Mining and Exploration Companies spent a little more than $5 million on 'broadcast of political matter', plus another $1 million on the publication of political material. Clive Palmer donated almost $1 million to the Liberal and National parties, with his Queensland Nickel company emerging as the largest individual donor to the Liberal Party.

By contrast, the Australian Labor Party, which has introduced a watered-down Mineral Resource Rent Tax, received no funding at all from mining interests according to the Electoral Commission.

Washington H. Soul Pattinson is a major donor to the Liberal Party, tipping in $250,000 in the 2010 and 2011 years, but it doesn't donate to get favourable treatment for its chain of chemist shops that includes Priceline. The company is a 60 per cent shareholder in New Hope Corporation, the developer of the New Acland coalmine on the Darling Downs.

The incoming Queensland premier Campbell Newman ostensibly rejected New Hope's ambitious plans for Stage 3 of the Acland mine, which would obliterate the town and more than double annual production to 10 million tonnes. Newman said it would be 'inappropriate' to expand the mine in the state's southern food bowl, and that he would oppose Ambre Energy's $3 billion coal-to-liquids project on prime farmland at Felton, also on the Darling Downs.[111] But even after these statements, the companies were optimistic. A New Hope spokesperson said: 'This has been the LNP's stated position during the election campaign and we will continue to work with the new LNP government.' For his part, Ambre Energy's director, Michael van Baarle, said his company would not be dissuaded from pursuing its project: 'We haven't had an opportunity to speak to the new government or the new bureaucracy.'[112] In fact, just five days after the election, Ambre was seen sinking more exploration wells on its lease.

Profitable miners have moved into the media and cultural sponsorship in a big way, and in some parts of Australia their logos are everywhere to be seen. It is hard to find a cultural event in Western Australia not sponsored by the mining industry. Park benches, schools, hospitals and even writers' festivals bear the imprint of mining largesse. The miners have moved into the film industry, having sponsored the film *Red Dog*, about a popular red kelpie that lived in Karratha in the 1970s. Next is sports sponsorship. Nathan Tinkler owns the Newcastle Jets soccer team and the Newcastle Knights league team, while five-times billionaire Clive Palmer wants to start his own national soccer league.

Gina Rinehart isn't content merely to own media companies; she wants to influence their coverage. After she took a seat on the Ten Network board, the network introduced a program by a prominent right-wing opinion writer and climate change denier. Now she wants to do something similar at Fairfax Media, which owns two of Australia's oldest mastheads. But the industry doesn't need to own the media to influence it, as shown by its ubiquitous television and print media ads. On the eve of the May 2012 Budget, the Minerals Council ran full-page advertisements opposing any changes to taxes that might affect it, even though such changes had not even been mooted.

The pressure applied to journalists and editors who run even mildly critical commentary is immense. A prominent columnist who rarely writes about mining says he is 'stalked' by an industry PR man who has sent him a series of harassing emails. The threats made to the ABC after it launched in late 2011 an information website about CSG shows how extreme pressure is applied. Rick Wilkinson, who runs the Australian Petroleum Production and Exploration Association, accused the ABC of using biased and emotive language because it used words such as 'toxic' to describe the chemicals used in fracking. He made a formal complaint to the ABC managing director and insisted on a series of factual corrections, such as a graphic that showed a drilling rig above a well without mentioning that the rig would be taken away when the drilling was done. Wilkinson also claimed that the ABC's figure of 40,000 CSG wells in Queensland was wrong, even though this number had already been confirmed by the state government.

When put together, these networks, lobbyists and sponsorship chequebooks represent a formidable influence, arguably giving the resources sector more clout than any other in this country. Unlike the farmers, miners don't need to set up their own political party to represent their interests; they already act as if they own the country. It is a case of a sector representing about one-tenth of our economy behaving like the tail that wags the dog. The level of influence is already significant, if not overwhelming. And if this is the extent of it when the industry represents about one-tenth of our economy, what will it be like if it controls a fifth or a quarter?

BOFFIN POWER

The resources companies' links with universities are one facet of their engagement that they are less inclined to promote. Although Australia's universities have become significant generators of export earnings in their own right, poor levels of government funding have forced them to be constantly on the lookout for money, and the miners have responded with significant funding but surprisingly little fanfare. The situation in Australia might reflect lessons learnt from the United States, where the gas industry lobby has openly courted universities as a way of addressing its poor public image.

Anthony Ingraffea, an engineering professor at Cornell University, says the industry has been buying expert opinion there in order to promulgate myths about its safety and impact on water and the environment.[113] He quotes Tisha Conoly-Schuller, the president and chief executive officer of the Colorado Oil & Gas Association, as an example of a deliberate strategy. 'The public is skeptical of anything we say,' she says, so the industry should get 'other messengers to carry positive messages about oil and gas to a skeptical public'. She argues that university professors are the ideal carriers, as they 'polled highest and are well-positioned in that regard'.[114]

Australian universities, including the Group of Eight universities who pride themselves as being world-class institutions, have been reluctant to reveal the extent of their funding by the resources sector. All eight were questioned about funding arrangements with resources companies, including those that prevent publication of research, and none was able to provide any details of this source of funding. The University of Melbourne declined to answer any such questions. A spokeswoman directed me to the university's Freedom of Information processes, which exclude documents that relate to 'matters of a business, commercial or financial nature'. The University of Sydney declined to provide any details. A spokesman said he had been advised that 'the names of specific companies, the value of the contracts/ royalties received from particular deals and details of terms agreed in relation to IP generally constitute confidential information. Sometimes even the existence of a contract is confidential.'[115] The University of Queensland said that any agreements with resources companies that gave the intellectual property to the company were classified as 'commercial-in-confidence information and it would not be appropriate for it to be published in the public domain'.[116] Even though the University of Queensland's Professor Chris Moran told me that the industry funds nine professorial chairs, the university would not officially confirm this. The University of Western Australia, which is believed to receive significant industry funding, did not reply to requests for information.

This poor record of disclosure is made worse by the fact that these Australian universities run courses for public servants from developing countries that preach the benefits of full disclosure when dealing with resources companies. In October 2011, Prime Minister Gillard launched the 'Mining for Development Initiative', which is aimed at promoting good governance in resource-rich developing nations. With $31 million in federal funding, the universities of Queensland and Western Australia will run courses for almost 2,000 officials from developing countries.

According to AusAID, Australia's overseas aid agency, a key principle of the Mining for Development Initiative is to encourage implementation of the Extractive Industries Transparency Initiative, a global anti-corruption program launched by Tony Blair in 2002 that involves the use of dual accounts to show all of the transactions between government and industry. And yet in Australia, government-supported universities won't reveal the full extent of their own financial connections with the mining industry.

One of the major beneficiaries of resources industry funding is the University of Queensland's Sustainable Minerals Institute (SMI), which has recently been obliged to reveal that as much as 90 per cent of its funding comes from corporate contributions. This level of funding does not mean that industry can 'buy' results, as the SMI's director, Professor Chris Moran, points out, but it does mean companies can influence what gets researched.

The Queensland University of Technology professor Kerry Carrington, who has researched fly-in, fly-out work practices extensively, points out, for instance, that the Sustainable Minerals Institute has not undertaken any research into the effects of these practices for almost a decade, and even then its research was not subject to peer review. There are also concerns that some of what SMI produces doesn't get published because companies control the intellectual property as part of the funding arrangement. Carrington spoke to SMI staff at a 2010 conference who complained that these arrangements meant their research 'never sees the light of day', and that this was undermining their academic advancement, which is based on the amount of material published in journals and elsewhere.[117]

SMI is a sizeable entity, with more than 300 research staff and an annual budget of about $30 million. While it has made some improvements in its public reporting as a result of media exposure, it is still well behind best practice. Up until late 2011, SMI had not produced an annual report since 2006, and all of its past reports had only disclosed a general total amount of corporate funding. After I reported this in an article in the *Australian*'s 'Higher Education' supplement, the University of Queensland's vice-chancellor, Paul Greenfield, demanded that SMI release more detail about its funding, including the amount provided by individual companies.

This new information showed that in the three years to 2010, about half of SMI's industry funding came from four multinationals: Anglo American, BHP Billiton, Rio Tinto and Xstrata. Industry funding ranged from 60 to 80 per cent over these years, but these figures don't include SMI's commercial entity. All up, total industry revenue is as high as 90 per cent.

The CSG boom has prompted SMI to expand even further, creating its latest subsidiary dedicated to producing tailor-made research for the industry. Launched by the then premier, Anna Bligh, in December 2011, the Centre for Coal Seam Gas (CCSG) is, as its name indicates, a centre *for* CSG.

While its mission statement is dressed up with rhetoric about engaging the community, it clearly sees itself as a driver of development. CCSG receives most of its funding from industry, although the details are not disclosed publicly. When the centre was launched, it secured funding of $20 million from the University of Queensland and three of the big four gas companies in the state: QGC, Santos and Arrow Energy. QGC, the subsidiary of UK giant BG Group, will provide half this amount, the university will provide a quarter, while the remaining $5 million will be funded equally by Santos and Arrow (the Shell–PetroChina joint venture).

For several months after the launch, the extent of the governance arrangements was covered in a single paragraph on CCSG's website, which states that CCSG will be governed by a 'Strategic Advisory Board' comprised of company representatives and chaired by the University of Queensland's vice-chancellor, or a nominee. Only after protests at the university about the new centre did SMI disclose on its website that the board representation reflects the amount of money contributed. SMI says the board provides 'advice' on the direction of CCSG and that the funding is 'untied', but there is no legally binding agreement underpinning these arrangements to protect research integrity. Professor Moran declined to answer a series of questions about CCSG's governance and that of the Sustainable Minerals Institute generally.

Origin Energy was a notable absentee from CCSG simply because its consortium, APLNG, is part of a five-year, $14 million partnership with CSIRO to create what is called the Gas Industry Social & Environmental Research Alliance

(GISERA). While the initiative reflects CSIRO's enthusiasm for being part of this new industry, GISERA is subject to transparent governance arrangements that ensure the corporate donor does not dictate or control its research, even though it has provided three-quarters of the funding. Decisions about research are managed jointly. An 'Alliance Agreement' ensures that all research is made publicly available, with research governed by an advisory committee comprising two representatives from each party. The contrast between CCSG and GISERA is stark, and reveals the extent to which universities have buckled to pressure from resource interests.

FIGHTBACK

The rains that broke the long drought in early 2011 in eastern Australia revitalised rural economies and brought new hope to regions that had suffered decades of devastating decline. The abundant new life on the grain fields and grazing paddocks west of the Great Dividing Range is there for all to see – a rural revival is underway. These rains came just in time for many regional centres faced with the prospect of mining and CSG developments on a massive scale. Instead of taking it lying down, these communities have rallied to fight for their future. While some, especially those with older populations, were worn down and beaten during the long drought, in others the younger folk are fighting back with gusto.

Much of the media attention has focused on those who believe that the only way to stop mining is through denying access to their land. This sentiment has driven the growth of the Lock the Gate movement, which represents a marriage of sorts between greens and farmers. Provided the two groups don't talk about climate change, the relationship is shaping up as a formidable force. Founded by Drew Hutton, who launched the Greens Party in Queensland, Lock the Gate has also gained support from the influential radio host Alan Jones, some undisclosed philanthropic backing and thousands of ordinary farmers around the country, who say they will lock their gates to prevent access to CSG developments.

Hutton has shown what is required to stem the tide of development. In December 2011 he was convicted of obstructing the QGC gas company and fined $2,000. But the case proved to be of some benefit to the anti-CSG movement because it revealed the close relationship between the former Labor government and the gas companies. The police who arrested Hutton were driven to the site in a QGC vehicle. Drew Hutton says that Lock the Gate has so far succeeded in preventing CSG companies from signing up more land-access agreements with farmers. He estimates that the companies have only 60 per cent of the agreements that they need.

Tim Duddy has shown that farmers need to work the political system as well as resort to the blockade. Farmers aren't able to stop miners doing exploration work on their land, but from 2008 onwards the people of Caroona staged a 615-day blockade against BHP Billiton's right to drill exploration bores on Duddy's farm. The farmers even adapted the yellow triangles of the successful Franklin Dam blockade with the words 'No Mines' replacing 'No Dams'. The blockade was followed by court action and a 2010 Supreme Court ruling against BHP. Duddy, a sixth generation farmer, then led an October 2011 blockade against plans by Santos to drill exploration wells nearby at Spring Ridge ahead of analysis of the region's aquifers by the Namoi Water Study. After a 20-day stand-off, Santos abandoned its drilling plans.

The battleground between farmers and miners in some ways resembles that of a conventional army under attack from lightly armed insurgents. Like guerrilla fighters, Australian farmers believe wholeheartedly in their cause and in the intrinsic value of what they do as food producers. But the potential loss of land value has also created a large element of self-interest that could prove to be a stronger and more enduring motivator for the individuals and communities locked in this epic struggle.

BLACK CAPITAL

The Indigenous affairs minister, Jenny Macklin, is a highly experienced political operator. Before entering federal parliament in 1996, Macklin led 'major strategic reviews' of federal health policy and urban and regional development policy for the Hawke–Keating governments, before in 2007 achieving her ultimate goal of running the vast social policy portfolio of Families, Housing, Community Services and Indigenous Affairs. The Indigenous part of her portfolio is one of the most challenging, and Macklin has focused attention on 'the gap' in Indigenous quality of life as compared to other Australians. In 2010, she embarked on national consultations to develop policies to address the crippling disadvantage that is the lot of most Indigenous people. But when she unveiled the Indigenous Economic Development Strategy in October 2011, what should have been a landmark announcement about economic empowerment became a long-winded advertisement for the mining industry.

The media responded enthusiastically to the launch of the new policy, but when they arrived en masse at Sydney's Sheraton Hotel, Macklin was surrounded by a phalanx of mining company executives who wanted to talk at length about their record of employing Indigenous people. The

Minerals Council of Australia (MCA), which represents the big resources multinationals operating in Australia, launched its own policy document, also called an Indigenous Economic Development Strategy, and proceeded to dominate the event to the point where the key message seemed to be that mining is the saviour of Indigenous people. The MCA's chairman, Peter Johnston, who runs Minara Resources, and its executive director, Mitch Hooke, outlined their industry's achievements in Indigenous employment. Then the diminutive figure of Jenny Macklin stepped alongside a tall MCA banner to announce the new policy of the federal government in this important area. Macklin had been completely blindsided by a carefully orchestrated PR campaign, in the same way that the Rudd government was totally outfoxed by the mining industry's advertising blitz against the resource super-profits tax in 2010.

That a federal minister would allow a single industry representing one-tenth of our economy to orchestrate the launch of a significant and broad-ranging policy statement speaks volumes about the power and influence of this industry over government and society as a whole.

Certainly, the mining industry has, albeit from a very low base, greatly expanded its Indigenous employment, though it is unable to provide exact numbers. Meeting Indigenous people who have moved from long-term unemployment into a well-paid mining job is an uplifting experience. Many of these workers – like the mother of four Lisa Brown, who works at the Boggabri coal mine – have been recruited off the street by mobile job-placement booths set up by mining companies.[118] Family networks have also played a role in helping Indigenous people break out of long-term poverty and unemployment: several interviewed at OZ Minerals' copper mine in remote South Australia said they were related to other workers at the mine.

But the achievements can be over-sold, as the event with Macklin suggests, and few analysts who have bought the story have looked closely at what lies behind the figures. In fact, there is little independent data on the amount of Indigenous employment in mining. Even academic experts rely on unverified data supplied by companies. This expansion has transformed the lives of many Indigenous people for the better, but mining itself is unlikely to substantially address the profound disadvantage of Indigenous people in this country for a very simple reason: mining is a capital-intensive industry that employs about 2 per cent of the Australian workforce.

In part, the industry's focus on employing Indigenous workers simply reflects the pressures of labour shortage. The federal resources minister, Martin Ferguson, suggests pragmatic rather than altruistic reasons for the shift towards greater Indigenous employment, saying that for many years the companies had 'neglected' a pool of workers living in the shadow of their mining operations.[119] But the emphasis on Indigenous employment also reflects the fact that about 60 per cent of Australian mining operations neighbour Indigenous communities, or are in fact on Indigenous land, so these employment opportunities are part of what's known as the 'social licence to operate' – the notion that companies benefiting from exploiting natural resources should make a significant contribution to the community. And in turn, mining companies want payback in the form of positive public relations.

Industry efforts have changed the views of some skeptics, including one of Australia's most prolific and best-known Indigenous academics. In 2010, the University of Melbourne professor Marcia Langton, a descendent of the Yiman and Bidjara people, argued that the 'resource curse' was being visited upon Indigenous people living on the edge of mammoth mining operations around Australia. To make the point, she looked at what they had done for the people living in the midst of the enormous development centred on the Pilbara town of Karratha, which services the Rio Tinto iron ore operations and Woodside's LNG plant.

Drawing on her account of a 2005 visit to Karratha and the nearby and largely Indigenous town of Roebourne, population 1,100, Langton said the mining boom was increasing the disadvantage faced by Indigenous people, not reducing it:

> The disparity between these towns is accelerating, and it is driven by the mining boom. In Karratha, everyone who wants to work has a job … In Roebourne, few people have the skills and education to join the fast-paced industries transforming the area.[120]

Langton added that even though Indigenous employment had been increasing, this trend was overwhelmed by the persistence of poverty on the edge of sprawling mining centres around the country: 'The contrast of Aboriginal poverty on the edges of the mining towns with the wealth of the mining workforce, despite the increasing numbers of Aboriginal people entering that workforce, is stark.'[121]

Two years later, Langton lauded mining as a major solution to Indigenous disadvantage. In an article written to coincide with Australia Day, 2012, Langton and Professor Matthew Gray from the Australian National University's Centre for Aboriginal Economic Policy Research (CAEPR) wrote that a 'quiet revolution' is taking place in Australia, as companies like Rio Tinto, FMG and BHP have created 'the largest indigenous industrial workforce in Australian history':

> In Australia, mining offers many indigenous populations a significant source of employment and contracting opportunities as an alternative to the welfare transfers on which many remote and regional Aboriginal communities depend. Rio Tinto Iron Ore and Fortescue awarded more than $300 million last financial year to indigenous contracting companies in the Pilbara. It was estimated last year that there were 52 contracting companies owned by indigenous businesses, or involved in joint ventures with indigenous companies. These companies are also employing indigenous people at an unprecedented rate.[122]

The article did not provide any national figures to support this claim, although it cited 1,500 Indigenous jobs at Rio Tinto and 300 at FMG, equivalent to 8 per cent of their workforces. These are numbers that would almost certainly be overshadowed by the livestock industry's employment of Indigenous people in past decades, albeit on lower wages.

Langton and Gray didn't cite any evidence to support their claim about employment in Indigenous contracting firms. In many cases, these businesses are run almost exclusively by non-Indigenous workers and managers. The authors qualified their positive appraisal, saying that 'high levels of disadvantage remain, even among indigenous communities located near resource projects', and that there is a danger with 'high levels of dependency on mining'.

The success claimed by the mining industry needs to be qualified in several ways. First, the increase in employment largely involves work at the bottom end, and this means that if another downturn arrives, it will most likely be a case of last on, first off, as has been the case in previous resources booms. Professor Jon Altman, who founded CAEPR two decades ago, argues that the sharp rise in Indigenous unemployment in 2009 in Western Australia and Queensland reflected the downturn in the mining industry. He pointed out that in the period immediately after the global financial crisis, 'some of the worst outcomes' were evident in resource-rich Queensland and Western Australia, where the Indigenous unemployment rate exploded to 20.8 per cent and 20.7 per cent respectively in 2009, from a more modest 12.7 per cent and 11.1 per cent respectively in 2008. 'In the space of a year, Western Australia went from having one of the best indigenous unemployment rates to one of the worst,' Altman noted.[123] In fact, this episode is almost identical to the trend observed after the early 1980s resources boom, when mining companies laid off large numbers of Indigenous workers.[124]

Second, the gruelling FIFO regime, with its two weeks of 12-hour shifts and long spells away from family and community, is a difficult routine for many Indigenous people, as indicated by apparent high turnover rates among Indigenous workers in the industry.

Third, many of these low-level jobs are now being filled by temporary migrants or are in fact becoming automated. The Enterprise Migration Agreement approved by the federal government for Gina Rinehart's Roy Hill iron ore mine will be the first of many such agreements that allow large numbers of workers to be put onto projects that involve an investment of more than $2 billion. Rio Tinto, the company that led the way on Indigenous employment, is now leading the way on the automation of everything from trains to trucks and diggers.

Finally, very little is known about the employment experiences of Indigenous workers, who are rarely interviewed in confidence. When mining companies grant access to their Indigenous workers, there is often a media person sitting alongside them. But some workers interviewed privately for this book say they have experienced discrimination and racism from white management and co-workers. One employee with more than 10 years' experience said there was little opportunity for advancement for Indigenous workers at his mine:

It is politics, I suppose. From what I am seeing here, there will never be an Aboriginal supervisor in the pit. My cousin had a meeting with the HR person about this stuff. I don't think she got back to him. You will never see an Aboriginal supervisor. They [Aboriginal people] are not getting anywhere. New people come in, never been on a mine site, they are on a different machine after 3–5 months. What is going on? Why are we in a truck when we have been here 2–3 years? There are two rules, blackfella rules and whitefella rules.[125]

The emphasis on mainstream employment for Indigenous people living in remote and regional Australia contrasts with the 'hybrid economy' model advocated by Professor Altman, which is focused on employment that reflects the Indigenous engagement with both tradition and the market. This hybrid model has been successfully developed by Galarrwuy Yunupingu on the Gove Peninsula in East Arnhem Land, where he has built enterprises that employ casual Indigenous workers in managing timber plantations, harvesting, sawmilling, construction and furniture-making. These businesses have designed jobs around the Yolngu people of the Peninsula, who retain many aspects of custom, especially strong ties to kin, in their social norms and everyday practices.

A key element in the success of these enterprises is the pride generated by Indigenous ownership and management. As one young employee says: 'These businesses are *ours*.'[126] The employment numbers are still modest – about 40 so far – but Yunupingu has done far better than the giant Alcan operation on the Peninsula, which has consistently failed to attract Indigenous workers over the past four decades.

There are many more examples of Gove-style success, most notably among outback communities that are generating carbon credits through controlled burning in remote areas. With the carbon tax in place, this is the perfect business model for Indigenous Australians – it allows them to connect with their country while generating sustainable economic enterprise. The former stockman Jack Green, a Garawa man from the Gulf of Carpentaria, has rejected offers to work at Xstrata's nearby McArthur River Mine and instead helped to develop a carbon abatement program that has substantially reduced intense fires during the dry season. The Waanyi/Garawa Rangers burn a 20,000-kilometre area of the Gulf country early in the dry season through aerial drops of incendiaries. They have set up 53 monitoring stations to track how their burning is reducing greenhouse gas emissions, and so far have reduced these by 63 per cent.[127]

Professor Langton's optimism about Aboriginal mining employment coincides with her joining the board of the Australian Employment Covenant (AEC), an initiative launched in 2007 by FMG's chairman Andrew 'Twiggy' Forrest. AEC originally set out to generate 50,000 jobs for Indigenous people, an ambitious target which after two years was replaced with the objective of generating 'commitments' by companies to employ Indigenous people. AEC was paid $4 million by the federal government, money that was collected by invoicing for each non-binding commitment that it generated. Figures released by the government show that AEC had placed 7,000 Aborigines into jobs, and only 2,100 remained in those jobs after six months as of December 2011. AEC contested these numbers and in May 2012 released new figures that showed 10,700 placements and a retention rate of 71 per cent. But these higher numbers exclude people who have been employed for less than six months. In total, 3,500 workers had been retained beyond six months. This is not an insignificant number, although it is a tiny fraction of Forrest's promise of 50,000 jobs. And as Dr Kirrily Jordan of CAEPR has pointed out, AEC does not provide any data to support its claim that all of these employees have come off welfare. Many of these placements could be of people who previously held jobs.[128]

Professor Langton accompanied Forrest when he addressed the National Press Club in May 2012. At the time she dismissed the figures released by the federal government about AEC's results. 'That's very dodgy – that's a very shifty thing to do with the figures,' she said. Asked if she was now working as an FMG consultant on Indigenous employment, Langton said, 'I am not directly contracted by FMG.'[129] Forrest has attracted some of the most influential Indigenous leaders to the board of AEC. Joining Langton is Noel Pearson, the director of the Cape York Institute, and Warren Mundine, the chief executive of NTS Corp, a business that assists Indigenous groups with native title claims. NTS Corp has been behind some ambitious land claims by Indigenous groups that are enthusiastic about mining. Some of these claims, including one by the Kamilaroi Nation that extends all the way from southern Queensland to the Hunter Valley, covers some highly valuable territory, but it also overlaps with the land claimed by the Wonnarua Nation, who are the traditional owners of the Hunter Valley. The Wonnarua people see the claim as an act of aggression. A Wonnarua elder, Victor Perry, says that, 'NTS Corporation is actively supporting land grabs by ambitious Indigenous groups across NSW.' Not only are the Wonnarua people having to grapple with the expansion plans of coal mines in the region, says Perry, but they're up against other Indigenous people: 'If we were back in tribal days, we'd be at war.'[130]

As for Roebourne, Marcia Langton's lament for the Aboriginal people living there might run even deeper if she returned today. The mining boom in nearby Karratha has bypassed the town made infamous by its gaol, and locals argue that the problems stem from a deliberate attempt by FMG to divide and conquer the community. FMG has given undisclosed funding to a breakaway group that is willing to accept a very low offer to mine iron ore on land claimed by the local Yindjibarndi people, who now mainly reside in Roebourne. The breakaway group, known as the Wirlu-Murra, has been lavishly funded by FMG to supplant the Yindjibarndi Aboriginal Corporation (YAC), which represented the Yindjibarndi people in their successful 2003 native title determination, and has the majority of applicants for an adjoining claim.

The Wirlu-Murra now has its own office in Sholl St, purchased for $600,000. They've also spent almost $1 million on legal actions, all aimed at destroying YAC, and about $500,000 on two well-paid consultants. Michael Gallagher and Bruce Thomas regularly work out of the office on a FIFO basis. Gallagher was a long-term FMG employee until 2010, when he left to take up a contract to advise the newly created Wirlu-Murra. Despite all these expenses, the accounts filed by the Wirlu-Murra with federal authorities show no income or expenses in 2010–11.

A rare insider's view of FMG strategy has emerged in an account given by the lawyer Kerry Savas, who worked for about a year as a legal adviser to the Wirlu-Murra. Savas signed a March 2012 affidavit filed with the Federal Court in which he said that FMG, 'through its employees and agent Mr Gallagher are directing the mind and will of the WMYAC [Wirlu-Murra]'.[131] Savas said in a separate interview that 'the court action they have put in place is an effort to wind up and destroy YAC'. The Wirlu-Murra is being funded to mount a Supreme Court challenge to appoint an administrator to YAC, while a Federal Court action is aimed at replacing YAC's native title claimants with those from the Wirlu-Murra. Savas says the strategy is designed to 'get the land for nothing'. He added: 'It is just about money.'[132]

Like every other mining company operating in Australia, FMG is not required to declare the handouts and funding that it has thrown at trying to destroy YAC. Like every other company, FMG keeps the results of negotiations completely confidential, which clearly serves corporate rather than Indigenous interests.

Australia is yet to sign up to the global anti-corruption regime known as the Extractive Industries Transparency Initiative (EITI). EITI has 13 fully compliant member nations, including developed countries like Norway, as well as more than 30 countries that are implementing the regime. The US president, Barack Obama, has said that his country will adopt EITI 'so that taxpayers receive every dollar they're due from the extraction of natural resources'.[133] Such a regime would require companies in Australia to declare all payments to government authorities, including native title groups, but so far the federal government has only endorsed the roll-out of a 'pilot' study. Australia's refusal to join this initiative clearly reflects the influence of the mining lobby, given that many companies operating here are in fact members of EITI.

DREAMTIME, JUST-IN-TIME

For companies racing to catch historic highs in commodity prices, compressing the time taken from board approval to first production is fundamental for success, and no project demonstrates this more than Woodside's development of the Pluto gas field located 190 kilometres off the coast from its Dampier LNG operations in Western Australia. Discovered in 2005 and approved for development by the Woodside board in July 2007, the $15 billion LNG plant started production just four years and eight months later in March 2012. While Pluto has been beset by delays and cost overruns, it remains a stand-out example of the breakneck pace of resource development in this country.

Workers involved in building and maintaining Woodside's gargantuan Dampier operations race to and from the plant in their 4WDs and utes. If you drive at the 100-kilometre per hour speed limit on the road that links the plant to the main road, you should expect to be overtaken by numerous Woodside workers along the short route. The carcasses of wildlife beside the road – including those of endangered species – are pungent reminders of one of the many costs of Australia's development rush. Locals have asked the council and the company to reduce the speed limit to 80 kilometres per hour in the hope that this might reduce the road kill, but they've been ignored.

The Pluto plant is another piece in the industrial jigsaw enveloping the beauty of the Burrup Peninsula. Pluto joins Woodside's Karratha Gas Plant, Rio Tinto's iron ore export terminal and salt plant, Burrup Fertilisers' ammonia plant and an industrial complex that services these mammoth operations. The LNG plants are breathtakingly huge and complex. As one tattooed Woodside worker said in the top-less bar at Dampier's Mermaid Hotel: 'It's a monster – it's a fucking monster.'

The Burrup's rapid industrialisation takes on a tragic hue when the loss of Indigenous heritage is considered. While much damage was done in past decades, even greater damage has been done in recent times, most notably with the Pluto development, and there's more in prospect.

The Burrup is such a heritage-rich and geographically small place that this could have been avoided; the peninsula measures just five kilometres wide and stretches for 25 kilometres. Photographs don't do justice to the colours there, the iron-red and deep crimson boulders that form ridges and hills laced with spinifex grass; nor do they do justice to the dense clusters of chiselled works of rock art found all across the area. These 'petroglyphs', or motifs, produced by the Yaburara people over the past 30,000 years, are undoubtedly the richest concentration of Indigenous rock art in Australia, and quite possibly in the world. The former WA premier Carmen Lawrence likens the Burrup rock art to Britain's Stonehenge, the painted caves of Lascaux in France and Peru's Machu Picchu. The art is yet to be fully surveyed, but studies indicate that there are as many as one million motifs. When small areas have been surveyed intensively, around 5,000 works have been identified in an area the size of two football fields.

The intricate engravings represent one of the longest records of continuous civilisation found anywhere in the world; they depict the wildlife and people of the region over a period of 30,000 years. They include more than 20 engravings of thylacines, which became extinct on the Australian continent more than 3,000 years ago. The National Trust has described the Dampier Rock Art Precinct as 'one of the world's pre-eminent sites of recorded human evolution and a prehistoric university'.

In 2007, the then environment minister, Malcolm Turnbull, put the Burrup on the National Heritage List, though he specifically excluded the area set aside for the Pluto plant. But this listing has not stopped industrial expansion, which is in fact accelerating. Twenty-one square kilometres of the Burrup's 118 square kilometres is already being used for industry, but another 69 square kilometres has been gazetted by the WA government for industrial development. Future industrial areas include King Bay South, where a memorial to the 1868 massacre of local Indigenous people stands. On a hill overlooking the site are 96 'standing stones' placed by the Yaburara people, as though they created a memorial to their civilisation long before the arrival of white settlers led to their rapid demise.

Western Australia's premier, Colin Barnett, has long declared himself a friend of the Burrup's rock art. As minister for resources in 1996, he released a land-use plan endorsed by the government that restricted industry to 38 per cent of the peninsula. He argued that with these controls in place, the unique heritage of the area could be preserved. 'I am confident that the strategy has struck a workable balance between encouraging industrial growth and the need to ensure the preservation of its cultural significance,' said Barnett.[134] He later described the Burrup's rock art as 'Australia's most significant heritage site'. As a backbencher in 2006, he said he regretted the loss of as much as a quarter of the art.

However, as premier, Barnett has sanctioned greater development, even when new projects have been opposed by his own advisers. In October 2010, against the wishes of the WA Water Department, he announced that a $370 million desalination plant would be built on the Burrup. He argues that the rock art can coexist with the rapidly expanding industry, although the balance between industry and preservation has clearly swung in favour of the former. The state government appears to do little to protect Indigenous heritage in this area and across the state. The Department of Indigenous Affairs does not have an officer based on the Burrup, and its budget for ensuring that resources companies comply with their Indigenous heritage conditions is a mere $300,000 a year – a tiny fraction of the royalties collected in 2010–11.[135]

Dr Ken Mulvaney, who works for Rio Tinto as a heritage specialist, first surveyed the Burrup rock art for the Western Australian Museum in the 1980s. He has located valleys where motifs can be found every few metres. He is able to distinguish between the more recent engravings, dating back 200 to 5,000 years, and those that possibly date back as far as 30,000 years. When Woodside's first LNG plant was built in the 1980s, the proposed plant threatened about 9,000 pieces of rock art, according to Mulvaney. About 4,000 pieces were preserved in situ, and an additional 1,000 were put into a compound at Hearsons Cove. This meant that the plant wiped out 5,000 individual pieces – and subsequent development has not been so kind.

Mulvaney is saddened by the lack of any real attempt to save the Burrup, and he fears that much heritage will be lost, not only through development, but also due to neglect. Meanwhile, the WA government is yet to complete a management plan, and there are no restrictions on access into valleys rich with rock art, which are frequented by mine and gas workers who tour them in their 4WDs on their days off. In fact, many of the works are being vandalised as a result of unregulated access.

A Woodside spokeswoman said that the company designed the Pluto plant in consultation with Indigenous groups in the surrounding area and avoided 92 per cent of rock art engravings within the lease areas. It set up a relocation program to 'relocate 171 boulders with engravings from the development area to nearby natural settings'.[136] Asked about the company's speeding drivers and the community's demand for a lower speed limit, the spokeswoman said, 'Woodside expects all its employees to comply with speed limits and local road rules.'

While Pluto is being bedded down, Woodside has begun preliminary work on its next just-in-time project, the building of a $40 billion LNG processing plant for gas from the Browse field at James Price Point, 800 kilometres north of Dampier on the pristine Kimberley coastline near Broome.

Even though Woodside has elsewhere proposed a floating LNG plant to avoid having to build one in East Timor, no such solution has been proposed for the Browse field. Woodside has negotiated a $1.5 billion compensation package with the Kimberley Land Council, but many local Goolarabooloo people and residents of nearby Broome remain bitterly opposed to the project and the impact it will have on their heritage and way of life. In what seems to be the start of an all-too-familiar pattern, Woodside began in May 2012 making applications to the WA government for permission to destroy Indigenous sacred sites at James Price Point for the purpose of building pipelines to bring the gas onshore. Goolarabooloo elders say the area is their ancient ceremonial ground used for the initiation of young men.

PART 3: THE DAMAGE DONE

HOME AND AWAY

The Margaret River region south of Perth should be far removed from the frenzied activity of Western Australia's resources hubs 2,000 kilometres to the north. This fertile oasis in an arid state bigger than Western Europe has become nationally and internationally renowned for wine, dairy produce and spectacular beaches. By adding tourism to the mix, the region has created a recipe for a diversified and labour-intensive economy that is also sustainable. This transformation was achieved long before mining came to dominate Western Australia.

But Margaret River is *not* isolated from the mining and gas developments in the Pilbara and around Karratha. The resources boom is fast catching up with the region, changing its fortunes for better and for worse. Sprawling housing developments are shooting up on the edge of the Margaret River town and several other towns in the region, consuming dairy farms as enormous new homes are built for a transient workforce. From Margaret River, mineworkers can take a bus to the major regional centre of Busselton and then fly to their jobs. Developers have devised an irresistible combination for fly in, fly out workers that offers the best of both worlds: the high-paying job in the resources industry and the rural setting for the family home. They promote Margaret River as 'FIFO Paradise'.

Such investment must surely be a boon for local business, given that the miners are at the very least bringing money and people into the region. But the verdict of businesses in Margaret River is decidedly mixed. The empty shops in the town indicate that the mining boom hasn't been all good for the region, as the sky-high dollar drives tourists overseas, while the influx of FIFO workers seems to have divided the community.

One local businessman, Russell McKnight, has ploughed the profits from the gold mines he developed in the Kalgoorlie region over the past 30 years into building a tourist complex based on the largest conifer maze in the Southern Hemisphere. McKnight thinks the 'mushrooming' FIFO suburbs are diminishing the unique value of the region, consuming vast amounts of high-quality dairy country and creating a bubble economy that is unsustainable. He laments the loss of prime farmland especially. Instead of building FIFO suburbs in Margaret River, McKnight thinks the government and the industry should be using the boom to develop the northwest of the state:

> Karratha should be the Dallas of Australia. The industry should be run from up there. If it was 50 years ago, you would be turning that into a city. There's a huge opportunity to develop the northwest in a sustainable way – they have got gas as well as iron ore.[137]

The influx of FIFO workers also brings other problems, he says. There is a hollowing out of the community, as fathers are away for long spells. The local rural fire service, of which McKnight is a member, is struggling to find sufficient numbers of volunteers to fight fires around the expanding town.

Instead of building new towns, as they did until the mid-1980s, mining companies now fly workers in from major cities and regional towns. As many as 100,000 mining workers live this way, and most of the projects are being designed around FIFO workforces. From the company perspective, FIFO offers an efficient and highly flexible model that removes the need for huge investment in building towns that one day may become ghosts. But the toll on people, families and communities like Margaret River is considerable, though the full extent is not well researched. What little information there is indicates that the concentration of mostly male transient workers in worker villages is fanning an increase in violence, drug and alcohol abuse and prostitution, as well as high rates of marriage breakdown. Towns with large numbers of FIFO families also have social problems, because children experience long periods without both parents at home.

The soaring rents in mining centres like the Pilbara and Queensland's Bowen Basin have given companies a justification to shift more workers into FIFO arrangements. In both places, a typical three-bedroom home can rent for $2,500 or even $3,000 a week. But a Karratha accountant, Gary Slee, who has seen booms come and go during three decades living in the Pilbara, contends that the government has failed to develop affordable housing alternatives, such as the pre-fab cottages that can be installed for as little as $150,000 each. In response to his own staff shortages, Slee developed a worker village that actually looks like a village – it has no fence or security gate, and the cottages are spacious. By contrast, he says, the local and state governments are fixated on building what he describes as 'Canberra in the Pilbara': sprawling suburbs made up of McMansions with double garages. This type of development is slow and is the main reason why rents are still sky-high.[138]

But FIFO may be less harmful than a variant known as DIDO, or drive in, drive out. Instead of flying, workers commute by car from regional towns, typically driving two hours or more at the end of a long stint of 12-hour shifts. DIDO dominates in the Queensland coal mines, which are located about two to three hours from the coast. Workers in more remote mines are known to spend seven hours driving each way to work. Car accidents on roads linking the Queensland coal mines with coastal centres have risen in step with the boom, indicating a deadly problem that the companies and government have done little to address. A third category involves the workers who live in mining towns like those in the Hunter Valley and Bowen Basin and commute each day to and from the mines.

Figures from Queensland's Department of Transport and Main Roads show a steady rise in road accidents, especially those involving heavy vehicles, in the Bowen Basin and Darling Downs mining regions over the past decade. The Warrego Highway that links Brisbane to the coal and gas fields of the Darling Downs has seen the biggest increase of all. Annual casualties rose 35 per cent to 303 between 2004 and 2009, but the number involving accidents with heavy vehicles rose 60 per cent to 66 casualties.[139]

The country road that links the ever-expanding Bowen Basin coal mines to Mackay is known as the Peak Downs Highway, but locals think of it as the highway from hell. The boom has led to a steady increase in accidents along this single-lane road that is crammed with commuting mine-workers and heavy vehicles, especially fuel trucks that cart diesel to the energy-hungry mines. Total accidents have risen 35 per cent between 2004 and 2009, the latest available year, while the number of casualties from heavy vehicle accidents more than doubled between 2004 and 2008, but then dropped back in 2009 as activity in the region slowed in step with the global financial crisis.

A community action group, the Mackay-based Road Accident Action Group, has posted home-made signs along the road, reminding motorists of places where accidents have occurred, and showing the cause of each accident and whether it was fatal. Some stretches of road have signs every few hundred metres. A former policeman, Noel Lang, who has been a key member of the group, says several things are involved in this rising level of carnage. He highlights 12-hour shifts as one of the contributing factors behind fatigue-related accidents, but adds that industry practices such as allowing workers to drive after finishing a night shift contribute to the carnage. Another key factor is 'hot bedding': as a result of accommodation shortages, workers have to clear their room at the start of their last shift so that another worker can move in. Should workers feel fatigued at the end of their shift, they have nowhere to sleep.

An extreme case of DIDO is experienced by mineworkers in central Queensland. One former maintenance worker, Josh Talbot, would drive seven and a half hours from his coastal home to work at Xstrata's Oaky Creek mine in central Queensland, and then do the same drive at the end of long stint at the mine. He and his wife lived on the coast because there was no accommodation near the mine. The drive would wipe out two of his five days off. Talbot says that although he was earning a six-figure salary at the Xstrata mine, the money could not compensate for the commuting and the hazardous work environment. He says that he often saw road accidents while commuting, and that miners from Oaky Creek were involved in accidents every couple of months. He reveals that some workers breached company policy and drove home after completing a string of 12-hour shifts. Asked if he would go back to coal mining, Talbot says the work was 'wet and dirty' with 'lots of coal dust' and adds that after the birth of his first child he would not go back: 'It's a young single man's life. I couldn't do it again.'[140]

In Queensland, the road accident that killed Senior Constable Malcolm MacKenzie and a mineworker in October 2005, followed by another involving a commuting miner in February 2007, finally prompted a coronial inquest. Both accidents involved commuting mineworkers who worked for BMA, a joint venture between BHP and Mitsubishi. The mineworker who caused MacKenzie's accident showed clear signs of driver fatigue: Graham Brown had just completed four back-to-back shifts and had driven for two hours when his 4WD veered onto the wrong side of the road.[141] The second accident involved the mineworker Robert Wilson, who showed signs of extreme fatigue at the end of his shift before he got into his car to drive home.

In February 2011, Coroner Annette Hennessy made more than 20 recommendations to improve driver safety. Many of them were still being considered a year later by the former Labor government at the time of the state election.

BMA's conduct shows that companies seek to deny any responsibility for work arrangements that are causing greater numbers of road accidents and deaths. As Hennessy reported, the company called expert witnesses to challenge the conclusion of the investigating police officer that fatigue was a factor in the MacKenzie accident. The company sought to discredit the officer – an experienced accident investigator – by arguing that his assessment was biased because he was a friend of MacKenzie.

12 HOURS+

FIFO and DIDO workers often work what's called 'nine and five', not nine to five. Nine and five means nine days on and five off, although it is becoming more common for these stints to extend to 14 days on and seven off. Construction workers on resource projects can work as many as four weeks straight. The workers do 12-hour shifts and then sleep in small aluminium sheds known as dongas. As a group, such mining employees are known as non-resident workers.

The British industrial sociologist Kathy Parkes, who has studied FIFO in the North Sea oil industry, thinks Australia's shifts are much tougher and that the roster would fail to meet EU working standards. Now an academic at the University of Western Australia, Parkes says: 'The work ratio seems to be higher here – more work and less leave.'[142]

The 12-hour shifts combined with rosters of nine to 14 days that crept into the industry about two decades ago were pivotal in allowing FIFO and DIDO arrangements to flourish. Had the eight-hour day remained in force, companies would have needed to rotate three crews over a 24-hour cycle; the 12-hour shift means they only need to maintain two crews. The mines operate seven days a week, only stopping on Good Friday and Christmas Day, although in the rush to roll out CSG projects, the gas company QGC had crews working on Good Friday in 2012 despite complaints from locals living near the work sites. In the space of 15 years, the mining industry has rolled back a century of industrial relations progress by creating a business model that would impress Henry Ford.

The reality for many workers is that their shifts extend beyond 12 hours, which can lead to a cumulative sleep deficit. This is especially the case for workers who live a driving distance to the mines and do shorter stints of about four days. Shifts typically involve a half-hour handover at each end, which means 13 hours' work a day. If these workers live only an hour's drive from the mine, that adds up to 15 hours a day taken up by work and commuting. Leaving eight hours for sleeping, this means that such workers have only one hour a day to do something other than work, commute and sleep. But some DIDO workers drive for more than two hours a day, and managers at these mines often work even longer shifts and still drive to and from work, which means less sleep and greater driving risk.

These 12 hour–plus shifts have also eliminated the need for companies to house workers in nearby towns. In Queensland's Bowen Basin, the state government has rejected plans for new housing estates and has instead approved thousands of new donga camps in and around towns like Dysart and Moranbah. In negotiations for a new enterprise agreement covering BMA workers, the company wanted 'flexibility' to increase shifts to 15 hours a day.

A retired Hunter Valley coalminer says the 12-hour shifts and long commuting clearly took their toll on his workmates, leading to general ill-health, depression and even suicide. The miner spent two decades in the Hunter Valley coal industry, including 10 years at BHP's Mt Arthur mine near Muswellbrook, before retiring in 2007. He says that 12-hour shifts made the work in the mines seem even more 'repetitious and drawn-out'. These shifts meant that the men worked a 51-hour week and, because of rostering arrangements, could never take all of their annual leave each year. He points to the consumption of Coca-Cola as one indicator of how the men managed to get through their shifts. In a typical 12-hour shift, the 60-man crew would consume as many as 200 cans of the sugary soft drink. Others resorted to a range of drugs.[143]

The former miner thinks that the mining companies will move away from DIDO workers and bring in more FIFO workers who can work longer 'strings' of shifts: a longer stretch of shifts makes mining more efficient because it reduces the number of crew changeovers. He explains, as only a coalminer could: 'Each time you have a changeover crew, that puts production down. The pits change while you are away, so each new crew needs time to familiarise themselves with the coal face. With FIFO, such changeovers are dramatically reduced.'

This assessment is spot-on: new research shows that most of the mega-mines currently being built will rely on FIFO/DIDO workers. This includes the Clive Palmer and Gina Rinehart coal mines in Queensland. A new report on future resource projects by Deloitte Access Economics makes clear that by 2020, resources companies expect to have 80 per cent of their workers engaged on a FIFO/DIDO basis in the Bowen Basin and the Northwest, with 55 per cent in the Surat Basin.[144]

IN DENIAL

The FIFO/DIDO phenomenon has far-reaching social consequences for the individual workers, their families and communities, yet little is known about its impact. Kathy Parkes points out that, 'The UK Health and Safety executive paid for my research, but there doesn't seem to be the same interest here.'[145]

For the cities and towns located near mining regions, there's unmistakably more life and economic activity, even though it comes in the form of unbridled male energy, with workers away from their partners, wives and children. The Karratha region of Western Australia, with a permanent population of about 16,500, has no fewer than 23 liquor licences, with eight of them approved for 'special facilities', meaning the work camps.[146]

Coastal towns in Queensland that have large numbers of DIDO workers now have a distinctly male culture. Queensland's Mackay is no longer a quiet and conservative rural centre servicing the sugarcane industry that attracted large numbers of Italian and Maltese migrants. The booming Bowen Basin mines have spawned a pub in the main street that proudly boasts of its topless bar, while numerous nightclubs operate until the early hours of the morning, even on Sunday nights. It's now a town of fast cash and fast cars; the Mackay Holden dealer has for the past three years recorded the highest sales for HSV Commodores in regional Australia. A local ambulance driver says that Mackay is a party town on Friday and Saturday nights, as large numbers of cashed-up miners hit the local watering holes with predictable results.

Getting reliable data on the FIFO/DIDO trends isn't easy; in many cases, the companies bluntly refuse to divulge it. This was the experience of the Cobar Shire Council, in far western New South Wales, when it tried to obtain accurate data, according to the council adviser Angela Shepherd. When dealing with resources companies, councils find themselves in a bind, as the expansion of mining increases the use of services that have to be maintained, without bringing about any compensating lift in revenue. In Cobar, companies have bluntly refused to pay for the local services they use. The council tried negotiating with listed mining company Cobar Consolidated Resources to upgrade a road for its proposed silver mine, but to no avail. Shepherd said the council was unable to put additional levies on companies to pay for services because rates were capped by the state government. The council could negotiate payments during the approval process, but it was unable to do so for existing mines, she said.[147]

One of Australia's leading demographers, Bernard Salt, is also frustrated by the lack of reliable information. Despite having done extensive research in this field, he is yet to arrive at a definitive national number of FIFO/DIDO workers and criticises the Australian Bureau of Statistics for failing to count the category in the 2011 census. He lists 13 towns in Australia that have large work camps, and I am aware of at least another six.

Academics at the Queensland University of Technology (QUT) School of Law and Justice stumbled upon the FIFO phenomenon when they began researching violence in rural Australia. As the authors explained in their submission to a federal parliamentary inquiry into these work practices, they 'did not set out with the intent of investigating activities of the resources sector', but their analysis of data for 'injury, harm, violent related mortalities, injuries and morbidity, influenced our choice of research locations'. As a result, they 'serendipitously came to conduct extensive field research in mining communities in WA and Queensland that are undergoing rapid socio-demographic redefinition due to the growth in non-resident workforces. This is also how we chanced upon the risks associated with camp life.'[148]

The researchers, led by Professor Kerry Carrington, found that in one WA mining community, rates of violence had risen almost threefold since the beginning of the resources boom, to a level 2.3 times the state average. In a Queensland mining community at the forefront of the boom, the rate of offences against people had grown from 534 per 100,000 in 2001 to 2,315 per 100,000 in 2003 – a four-fold increase to a rate more than twice the state average.[149] Domestic violence had emerged as a notable problem in mining regions of Queensland. Breaches of domestic violence protection orders in 2008–09 for one such region were 1.63 times the rate of the state capital. In the SA mining town of Roxby Downs, which serves the giant Olympic Dam uranium mine, rates of recorded offences against people were, in 2006, 1.36 and 1.39 times greater than the respective state and regional averages.[150] Carrington points out that these figures most likely underestimate the actual amount of violence because offences are often not reported. One senior police officer who works in a major mining town told Carrington that most instances of violence are not reported, and the crime statistics are 'not worth the paper they're written on'.

Other research, as well as anecdotal evidence, indicates that the social implications of FIFO/DIDO are serious and in need of strong policy remedies. At least the politicians think so. This is why federal parliament's Standing Committee on Regional Affairs launched a wide-ranging inquiry into the practice, which has generated a wealth of new material from community sources, most of it negative. Even so, there still remains a lack of scientific research into the FIFO effect.

A WA-based clinical psychologist who cannot be named said that 10 per cent of her 400 clients were FIFO-families, although most of the people attending her clinic were the spouses, not the workers. Of this cohort, FIFO was a major contributor to the problems that brought these clients to the clinic. Problems included: a growing sense of distance within the family; a difficulty in adjusting to the different lifestyles of home and work; affairs with workmates; relationship break-down, often involving poor communication and arguments over distribution of workload during weeks off; and depression, anxiety and sleep disorders. When asked to rate the extent to which FIFO was a factor in the relationship break-downs, the psychologist gave an estimate of 90 per cent, although in most of these relationships there were often already problems that had been exacerbated by the FIFO work. She also qualified her criticisms: 'I see a very biased sample of the FIFO workers. I have some friends who are FIFO workers who think it is the best thing for their marriage; keeps it fresh and alive. Clearly I don't see any of these people.'

Another study led by Carrington at QUT reported evidence from the partners of mineworkers that points to serious concerns about alcohol and drug use, as well as money management. The partners' coping mechanism seems to be the knowledge that the mineworker will soon depart for another roster. In the words of one woman:

My partner works in the mining industry and stays out at camp during his working week. He has problems with drinking, drug use and money management. He drinks

until he is drunk most nights out at camp and comes
home and wants to do the same on the weekends. I find
that him being out at camp for longer than he is home is
very detrimental for his health and also for our relationship … This is not an isolated relationship – most of the
women I meet and speak to about how they cope with
their partner being away in the mines have developed a
kind of coping mechanism where they have allowed
their partner to do as he wishes because he will be leaving to go back to work anyway.[151]

When Professor Carrington first reported these facts in
early 2011, the industry went into meltdown and used the
media to attack her credibility, labelling the report as 'dodgy'.
Michael Roche, the chief executive of Queensland's powerful
lobby group, the Queensland Resources Council, went all
the way to the vice-chancellor of QUT to complain about
the report. While the report's findings were one thing, what
concerned Roche most was that the university had issued a
press release to promote it. This, from a former PR man who
works for an industry that had just spent $22 million on an
advertising blitz that involved highly misleading statements
and played a significant part in the removal of Prime Minister
Kevin Rudd.

QUT buckled under the weight of Roche's assault. In a
classic case of shooting the messenger, senior management
responded a few months later by removing the media and
marketing manager who had put out the press release. Ian
Eckersley, a former veteran journalist with the ABC, suddenly

found his position made 'redundant' at the end of 2011. Early in the following year, QUT re-advertised the role.

In its submission to the federal parliamentary inquiry, the Queensland Resources Council repeated its claim that only 15 per cent of mineworkers in the Bowen Basin are non-resident, a figure strongly disputed by Carrington, whose estimate is 68 per cent. In a subsequent submission to the inquiry, Carrington argued that the 15 per cent figure grossly understates the FIFO/DIDO numbers because it excludes contractors and workers who are currently not rostered on. It also excludes workers living in the major coastal centres of Mackay and Yeppoon. Carrington concluded that her estimate is based on data from the state Development Department, which identified almost 15,000 FIFO/DIDO workers in June 2010 and which she adjusted for shift rotations.

Carrington's estimate accords with evidence from Western Australia, where 38,000 FIFO workers indicates a rate of 70 per cent in the mining workforce, according to a report by Heuris Partners for the Pilbara Industry's Community Council.[152] But the numbers could be even higher, because in free-wheeling Western Australia, mining companies just go ahead and build camps or expand existing ones without getting approval from the authorities. Hancock Prospecting, the private company owned by Gina Rinehart, built just such an unauthorised camp for 59 staff at its Roy Hill prospect in 2007. It was fined $20,000 for this violation. The fine obviously proved no disincentive because the company was at it again two years later. As the resources minister, Francis Logan, said at the time, the company had shown a 'blatant disregard of regulations and tenement conditions'. Logan made the point that while companies are urging government to speed up the exploration approvals process, some companies are showing a 'complete disregard for the environment and blatantly disregarding the exploration conditions'.[153]

The Queensland Resources Council figure of 15 per cent for Queensland was repeated in a submission by the heavily industry-funded Centre for Social Responsibility in Mining (CSRM), which is part of the Sustainable Minerals Institute. In a nine-page submission that included just three references, CSRM dismissed any concern about FIFO/DIDO work arrangements, claiming that 'FIFO and residential options offer different lifestyle solutions that appeal to different segments of the population'.

THE END OF COMMUNITY

The rugby league player Travis Norton looks back at the Queensland mining town where he grew up as an idyllic place where people worked hard but still had time for each other and the wider community. 'We moved to Moranbah when I was four or five and I lived there until I was 18, did all my schooling there. I don't think there could be a better place for kids to grow up,' he says. This was in the days when miners worked a normal eight-hour day, the dads were there after school to pick up their children, coach football teams and become involved in community life. Norton went on to captain the Queensland Cowboys and play in the State of Origin series.

The single biggest factor that has changed Norton's hometown for the worse is the move to the 12-hour shift. He says he has nothing against the industry – some of his brothers are mineworkers, following in the footsteps of their father. But Norton bitterly laments the change in working hours: 'Now everyone is on 12-hour shifts, so they are either working or sleeping and most of them have homes somewhere else anyway.'

For Moranbah and surrounding towns in the Bowen Basin, like Dysart and Middlemount, rotating work-shifts have turned a normal town into one being strangled by work camps. The work camps, or 'villages' as the companies like to call them, have their own gyms and bars, which means that workers don't have to spend their money in the towns. But the camps don't have their own clinics, meaning this transient workforce often overloads the local medical services.

The Moranbah Traders Association president, Peter Finlay, says the state government has not been interested in expanding Moranbah because there is coal adjacent to the edge of the town and it's covered by a mining lease. His proposal for a new residential development was rejected by the state government because it was concerned about losing the tax proceeds. When he met with the state opposition to seek support, they told him they didn't want to miss out on the coal royalties either.

The mining company BMA claims that it is investing in new housing, and some homes can be seen going up in a new subdivision, but the construction process is far too slow to meet demand. BMA won special approval from the former state Labor government to operate its next development,

Caval Ridge, with a 100 per cent FIFO workforce. The company is already rolling out work camps in the centre and on the outskirts of Moranbah like a military operation; it seems to be investing far more heavily in expanding the capacity of the local airport than in permanent housing. New recruits are being told that if they want a job, it must be on a FIFO basis. Says Finlay:

> FIFO is fantastic for the mining companies, great for Qantas, but I'm convinced that people don't want it. We're losing families every week who can't afford to stay here. Locals are being replaced by FIFO workers. It is killing the town. I am president of the traders association. They have not contacted me. I can't find any business that's earned money from them. It is a sham.[154]

Doctors at the Moranbah Medical Centre say the camps are breeding grounds for illness. In recent years there have been outbreaks of conjunctivitis, influenza, gastroenteritis and whooping cough caused by having large numbers of workers living in close quarters. The medical centre manager, Laura Terry, says that about a quarter of the hospital's patients are non-resident mineworkers, and one in three emergency patients are non-resident mineworkers, and yet this 'hidden population is not funded'. The centre is trying to convince the companies to fund accommodation for doctors looking to relocate to the region, but has so far had no success.

When the medical centre commissioned a study to examine the types of illnesses that caused mineworkers to seek outside medical help, it found that half the non-resident patients treated in the emergency ward were classified as 'inappropriate', meaning that they were suffering minor conditions that could have been treated by a GP. This indicates that the mining companies are relying on free local services rather than sending staff to local GPs or providing their own medical clinics.[155] BMA has recently started sending FIFO doctors to work at its Bowen Basin mines, says Terry, rather than responding to community demands for housing to help attract resident doctors. These doctors work in the mines and work camps only, and because they are not locally registered they're unable to perform basic tasks such as referring a patient for an x-ray. Terry says these doctors must be costing BMA a fortune, yet they are 'ineffective'.

While the FIFO/DIDO workers drain these essential services, locals say they are putting very little back. John Crooks, who owns the hardware store in Dysart, says the new camp being built on the edge of town will become a gated community, with its own golf course built alongside the public golf course. He thinks the camps are deliberately designed to ensure that workers aren't part of the town community. Crooks grew up on a farm, began his working life in the resources industry and has spent 25 years in Dysart. He says he gets a lot of business from the mines and is not anti-mining by any stretch, but he thinks that the companies are no longer interested in having people live in Dysart as permanent residents. Crooks and his wife Marina raised their three children in the town and they all went to the local school. He says the facilities are 'fantastic', but now under-utilised.

Donna Partington, a mother of six who works part-time in the Crooks' hardware store, has refused to take a job in the mining industry because she wants to be around for her children. She prides herself on her involvement in the Girl Guides, but she's seen dwindling numbers of people willing to work in community organisations. Partington knows of cases where both parents work in the mines and their children are left on their own for extended periods. She thinks that the work camps are killing the town. 'The older community is retiring and moving away. People are more interested in money, not family and community,' she says.

Instead of prospering from the biggest mining boom in 150 years, mining towns are being worn down and driven into the ground by the FIFO/DIDO phenomenon. Shops are closing in Dysart and Moranbah despite the huge influx of workers, and the towns and the people seem exhausted. Marina Crooks says the long shifts and all the driving make people physically tired to the point where they start to break down, especially the older men. John Crooks says the miners think they get a lot of time off in return, but 'they spend half of it driving'. The Crooks know all about the consequences of this combination of shift-work and driving too well. Tragically, their 19-year-old son Tristin died in a car accident on his way to work one morning in 2010. He was working at BMA's Norwich Park mine, which the company closed in April 2012 because it had become unprofitable.

NOWHERE TO HIDE

Around the old mining town of Acland on Queensland's Darling Downs, where underground mining has happily co-existed with farming for almost a century, a coal company, New Hope, has submitted a third expansion plan to the state government that would double production yet again, leading to the complete destruction of the town that was promised great things when the mine opened a decade ago. So confident is the company of getting the project approved that it has already bought up most of the weatherboard homes in the town and dumped them in a paddock, along with the school and surrounding properties. Only one permanent resident remains in a town that had a population of 200 before the mine began buying people out.

Nearby lives a grazier, Tanya Plant. At 36 years of age, she seems to have much to look forward to: a cattle farm that's been in her family for four generations and two young children. She obtained a doctorate from Oxford University on a Rhodes scholarship. But there is a thorn in her side: Plant's two-year-old daughter has been having coughing fits and her doctor thinks the cause is 'environmental': the dust particles, gases and metals generated by the mine that operates just two kilometres from her home. Her daughter also complains about the constant 'growling noise' she can hear, even at night. Plant has reached the point of thinking that her family can no longer stay on their land:

> I'm only 36 and had hoped and expected to continue to live an active life for some time yet and to be able to raise our kids in a good environment to give them the best start and chance in life. This farm has been in my family for many generations and is very much a part of us. I can't really picture a happy future without it but I'm not sure whether we should live here anymore.[156]

The mine has the ironic name of New Acland. The Environmental Authority under which it is allowed to operate prescribes dust levels that are three times the state guideline, while noise limits are a third higher. Even so, the mine often breaches these limits, but the government has so far ignored this.

While Plant is most concerned about her eldest daughter, she's worried about every member of her family: 'I'm uncomfortable telling too many people the details of all our health issues, but there are some worrying symptoms which seem to have been going on for quite a while and none of us seem as healthy as we should.' She's concerned about the effect of 'those really small particles, as well as things like heavy metals'. Despite persistent efforts, she's had little success in getting access to reliable information from New Hope Corporation, which is majority owned by Washington H. Soul Pattinson, which in turn owns a share of the pharmacy network that trades under the same name.

New Hope seems to have had a cosy relationship with the state government. When an expert from a government agency came to Plant's home to measure for dust, he was telephoned by a senior executive of New Hope, who arranged to meet him immediately after completing the test. In fact, the Plant family are yet to see the results of this monitoring because the company has made the release conditional on Tanya Plant submitting to yet another meeting with the executive. She is still demanding testing for the ultra-fine particles that particularly worry her.

The experience of people who have tried to stay on their land and battle the ill-effects of mining, and corporate power and government indifference, is not encouraging. Like Tanya Plant, Margaret Hassall and her family lived very close to the New Acland mine for about eight years. The pit came as close as 1.2 kilometres to their home during this time.

Margaret, her husband and their two daughters had moved to the Acland farm in 1991, where they began breeding alpacas. In 1998, Margaret's parents, Aileen and Ken Harrison, sold their farm and decided to build a brick home on the Hassalls' property, putting most of their savings into the venture.

In 2001, New Acland began operating as a small open-cut coal mine. Four years later, in 2005, it won approval for its 'Stage 2' expansion that would push output to 4 million tonnes a year. As the mine encroached on the farm, the family was inundated about twice a week with the effects of blasting, which often involved noxious orange clouds wafting over the property. Dust cloaked their homes and made its way inside, and then there was the noise, 'like having tractors on every side of the house 24 hours a day', says Margaret.[157] When the company blasted – often without warning – the entire house shook violently, then dust hung over the house for a day or more. The Harrisons could see the blasting from their verandah, the gases and dust burning their eyes, nose, throat and chest. The dust, noise and plumes of gases turned this rural paradise into something akin to a war zone.

New Hope engaged a contractor who carried out ad hoc testing for PM10 dust particles. Not only did he fail to test for all types of dust particles, but, Aileen says, he also chose the most favourable times to test, such as after it had rained, or when the wind was blowing away the dust from the farm. Margaret kept a diary during this time, and one of her entries after a blast says:

> It started the night before with the noise. It was so loud that it was really intrusive and I hardly got any sleep. The noise decreased mid-morning, just in time for the blast. This one was a big bang ... it seemed to be double-barrelled, so it lasted for several seconds. The windows shook, the internal walls shook, the furniture shook, and I could hear the shed outside shaking also. Then the dust started. As the day went on it got worse and worse. I had a mask on from about 3pm. I hope I survive the day, cause at the moment I don't feel like I will.[158]

The mine altered the health of the family for the worse. They all felt lethargic and one daughter had to take anti-histamines so she could sing at school. Aileen was quite severely affected: she had been in good health before moving to Acland, but she began suffering severe attacks of asthma-like symptoms. She collapsed on three occasions, and after one of them could not eat for 36 hours. She felt much better whenever she left home and went into town for a few hours.

The mine also ruined the family business, which had attracted a steady stream of tourist buses to look at their alpaca herd and buy goods from their shop. Once the mine was operating, the tourists would step out into the dust and then get straight back on the bus.

After several years, the family decided they'd had enough, and looked to sell up to the mine. But when it came to the financial settlement, they were astounded by the behaviour of the highly profitable company. Its strategy seemed to be one of wearing them down. On one occasion, when the family refused his low offer, a negotiator stormed out of their home after knocking over furniture.

The only bargaining chip the Hassalls had was an agreement reached prior to the mine commencing, which said that the company would agree to buy them out if any family members suffered ill-health related to the mine. The Hassalls are prevented from discussing because the company obliged them to sign a confidentiality agreement. However, Aileen Harrison signed no agreement, and she reveals that the terms offered by New Hope in no way compensated the extended family for its once valuable property. Unlike most rural properties, the farm had four land titles, which made the land more valuable because it could be subdivided and sold as four separate properties. This would have greatly increased the value if the mine was not there, but the settlement didn't take this into account, even though the company claims it paid above the market value.

Aileen and Ken, who should be enjoying a happy retirement after a lifetime of hard work, can now only afford to live in a hot and cramped aluminum demountable home that resembles a school classroom. Most of their possessions won't fit in their new abode, while their antique furniture looks sadly out of place.

After all of their difficulties, the Hassalls have found it impossible to get away from the miners. Five months after the family moved into their new property at Kingsthorpe, 30 kilometres from the Acland mine, they discovered that a new exploration lease covered their farm.

FROM PIT TO PORT

Like the impact of most mines, that of the New Acland mine extends far beyond the immediate fallout zone still populated by stoic farmers. From the mine, the coal is trucked on a local road to a coal dump located just one kilometre from the town of Jondaryan, population 377. Satellite images reveal that the sprawling dump measures 700 metres long and 300 metres wide, giving it a far greater expanse than Parliament House, Canberra. When asked why New Hope had located its coal dump so close to the town, a spokesman said the company was fulfilling all the legal requirements set by the state government.

The trucks are meant to be completely covered to prevent lumps of coal from flying off the back, but the roadside is littered with coal for the entire 10 kilometres from the mine to Jondaryan. The state coordinator-general was following the route in late 2011, as part of his assessment for the Stage 3 expansion, when a lump of coal hit the windscreen of the car he was travelling in.

From Jondaryan, some coal is regularly trucked through the main street of Toowoomba to local power stations, but most is sent by rail to export terminals in Brisbane. People who live along the railway lines, and in adjacent towns, say there has been dramatic growth in coal trains over the past decade and they now have black soot on their roofs that gets into their water. Peter Faulkner, who lives just 300 metres from the railway line, has black streaks on the plastic water tank he uses for his drinking water. Asked if he has considered obtaining an assessment from the government, Faulkner says he no longer trusts it: 'There's no impartiality when it comes to assessing these mining projects. The fact that they seem to be covering everything up concerns me greatly. They have a duty of care towards us. They are not looking after us at all.'[159]

While state governments insist trucks be covered with tarpaulins, the same rules don't apply to coal trains. The Queensland Department of Transport argues that as the 'coal is treated with water ... this helps bind dust particles to the surface of coal to reduce dust during transit'.[160] Meanwhile, NSW's RailCorp is leaving it up to the industry to act: 'Rail-Corp understands some freight operators are investigating ways to improve the wagon seals and we await the results with interest.'[161]

Like most mines in Australia, New Hope is not subject to mandatory public disclosure of its dust, noise and gas monitoring. The company is able to produce its own tests at the most favourable times, and the results that are made public are based on a rolling average, which erases the sharp peaks.

In a letter to the *Australian*, New Hope's managing director, Robert Neale, claimed that coal dust and its emissions were completely harmless. He cited a report on Gladstone, located 500 kilometres to the north, which concluded that there was 'no indication that coal dust is causing any likely risk to health' and that 'exposure to coal dust in itself is not posing a specific risk to human health'.[162] But as one reader responded the next day, the personal address given by Neale at Brookwater, renowned for being the state's 'premier golf community', was far, far away from the coal dust of Jondaryan and Gladstone. While Gladstone has heavy industry and power stations, it doesn't have coal mines and it certainly doesn't have a coal dump like the one at Jondaryan.

TOXIC TOWNS

The particles that concern people living near coal mines are microscopic. One of the smallest, which measures 2.5 microns or PM2.5, is one-thirtieth the width of a human hair and can penetrate deep into the lungs and enter the bloodstream. Some experts argue that there is no safe level for

PM2.5. The larger variety, known as PM10, measures one-seventh the width of a human hair. Both types of particles are generated by coal mining and by the burning of coal. So far there has been little measurement by government of these particles near mining regions, and the measurements of the companies cannot be relied on. Companies will sometimes measure PM10 levels without checking PM2.5 levels – this is what New Hope did on the Hassall and Plant properties.

In China, where air pollution has become a major problem, blogs are running hot with references to PM2.5. In Beijing, where it can be difficult to see through the smog to buildings across the street, the government began publishing readings for PM2.5 in January 2012. In March 2012, Premier Wen Jiabao pledged national standards by 2016, which would be combined with measures to reduce pollution.

By contrast, Australian governments pay scant regard to the impact of these microscopic particles, especially in mining regions. In New South Wales, a network of 14 dust monitors has been rolled out in the Hunter Valley in response to community demands, but only three of these measure PM2.5. In Queensland, the government proudly boasts of a network of 25 air-quality monitoring stations designed to give the public accurate and timely information on levels of dust particles and the toxins produced by burning coal, sulfur and nitrous dioxide. This reassures the public that the polluters are being watched day and night. But in fact only three of the government's monitoring stations are located in mining regions, and none of them measures the ultra-fine PM2.5 particles.

Few people in Queensland would be able to readily access on the State Department of Environment website the really important information concerning specific levels of arsenic, copper, lead and zinc in the air around each monitoring station, as there's no obvious path to these pages. And there is a further problem: this information can often show reassuringly low levels of pollution when in reality levels are dangerously high.

The case for good monitoring and emission controls in places near lead smelters could not be more compelling, given the conclusive research about the consequences for childhood brain development that led to the removal of lead from petrol in Australia and many other countries. But the regional and remote towns that host lead mines and smelters continue to be subjected to a toxic assault as governments look the other way. In South Australia's Port Pirie, the home of Swiss-based Nyrstar's lead smelter, the state health department found in a 2012 report that 29 per cent of children aged from birth to four years had blood lead rates of more than 10 micrograms per decilitre. This level is the Australian standard set by the National Health and Medical Research Council, which is in fact double the standard recently adopted in the United States. More than 6 per cent of children had rates above 15 micrograms.[163]

In Mt Isa, Swiss-based Xstrata operates the Mt Isa Mine, which has a smelting operation located just two kilometres from the town centre. Three chimney stacks belch the results of lead processing with impunity. The results of the lead mining and smelting operation that's been going on since 1931 can be found all over the town, even in the soil. The rate of children with lead blood levels above 10 micrograms per decilitre has ranged between 5 and 10 per cent in recent years. Professor Mark Taylor of Macquarie University argues that if the US standard was adopted, 50 per cent of children would have levels above it. Two children, Sidney Body and Bethany Sanders, recorded blood lead levels of 31.5 and 27 micrograms per decilitre respectively. They were aged four and five at the time these readings were taken in 2010.[164]

Located in the Gulf country 1,800 kilometres northwest of Brisbane, Mt Isa is so remote that the health problems of its 21,200 people seem a low priority and confusion reigns over the data provided by the government about the level of lead in the air. At first glance, levels throughout 2011 appear to be less than a third of the state 'guideline', which is stated as two micrograms per cubic metre of air – and so far in 2012, they are even lower. But these graphs are misleading because, according to the government's own *Environmental Protection (Air) Policy 2008*, the limit for lead is actually 0.5 micrograms per cubic metre. (It is worth noting, too, that this is three times the 0.15 micogram limit adopted by the United States.) Thus, the Queensland government's own data, which is based on average levels over a 24-hour period, shows that in fact the mine breached the official 0.5 microgram limit in seven out of 12 monthly readings in 2011.

A similarly confounding result arises with arsenic levels. The graph for a monitoring station at The Gap, 1.5 kilometres east of the Mt Isa mine, shows readings sitting well below the red line that marks the 'guideline' of 0.3 micrograms per cubic metre of air. But this limit is wrong; in fact, it is just six *nanograms*, which means the website is overstating the limit 50 times over, and the readings on the government website show that the Mt Isa operation is consistently exceeding this limit by a factor of 50. When converted from micrograms to the correct unit measurement of nanograms, the arsenic from this mine was 2,970 nanograms, or 495 times the state limit.

This complex data was interpreted by Mark Taylor, who believes that the state government is turning a blind eye to gross levels of dangerous pollution: 'They are misleading us in a systemic way. It is not only done by the mine but by the government. They are not even asleep at the wheel; they are looking the other way.' Taylor adds that where readings exceed legislated limits, they are likely to underestimate pollution levels because they are based on a 'rolling average' that 'makes it really difficult for the polluter to exceed the limit'.

When asked to explain these results, Andrew Buckley, a regional services director with the state Department of Environment, said the air-quality information on the department's website is not for compliance but rather to inform the Mt Isa community: 'As such not all of the parameters displayed on the webpage can be directly compared to compliance limits.'[165] Buckley explained that Mount Isa Mines Limited had, until late 2011, operated under its own special legislation from 1985, which meant that air-quality levels only had to meet the regulatory requirements under the *Mount Isa Mines Limited Agreement Act 1985*. Not until December 2011 were the general environmental limits for sulfur dioxide, lead, PM10, PM2.5, arsenic and cadmium applied. The company has now been asked to provide more data.

FROM DUST TO DUST

One of the problems with producing definitive results about the ill-effects of coal dust and emissions is Australia's relatively small population, especially in areas near mines and coal-fired power stations. Unlike in much of Europe and the United States, Australians have greater choice about where to live, and they try to avoid such places as much as possible. As a result, specialists in this field, known as epidemiologists, say that detailed studies have not been done in coal-mining regions here because they would not yield 'statistically significant' results.

In response to community concerns, the NSW government produced a study on the health of people in the Hunter. The data was based not on physical examinations, but on the data from hospital admissions and telephone surveys. The results were mixed, although they did show that people in Muswellbrook had higher rates of asthma.[166] Dr Tuan Au, a Vietnamese refugee now working as a GP in Singleton, has done what the government won't do and conducted physical examinations of children under the age of 15. This study of 683 school students found that the rate of below normal lung function was three times higher for students who lived near mines, compared with those living further away.[167] Speaking as a refugee, Tuan says he feels for the people of the Hunter Valley because many will be forced to leave their homes.

Studies on larger populations in the United States and Europe indicate that exposure to coal dust and the emissions caused by burning coal can produce a range of effects, all the way from asthma to heart disease, leading to premature death. A 2001 peer-reviewed study of 17,000 West Virginians found that people living in high coal-producing counties experienced significantly higher rates of cardiopulmonary disease, chronic lung disease, hypertension and kidney disease when compared to people living in low coal-producing counties.[168]

The study also found that mines producing more than 4 million tonnes a year – such as the Acland mine – had consistently worse effects on the health of nearby populations:

> As coal production increased, health status worsened, and rates of cardiopulmonary disease, lung disease, cardiovascular disease, diabetes, and kidney disease increased. Within larger disease categories, specific types of disease associated with coal production included chronic obstructive pulmonary disease (COPD), black lung disease, and hypertension. The highest level of mining (>4.0 million tons) predicted greater adjusted risk for cardiopulmonary disease, lung disease, hypertension, black lung disease, COPD, kidney disease, and poorer adjusted health status.

As with any study, there were limitations, such as the likelihood that the populations living close to mines had higher rates of smoking and that they may have worked in mines or related industries. But the authors argued that three of the five limitations may have actually led to under-reporting of the health effects. The authors contend that the threat to human health from coal mining stemmed from exposure to by-products, 'slurry holdings that leach toxins into drinking water and air pollution effects of coal mining and washing'.

Tanya Plant's concern about heavy metals may arise from her family's proximity to the coal washing facility, which uses polymer chemicals and large quantities of water. The process creates liquid waste called 'slurry' – with the consistency of wet cement – that must be stored.

A 2009 report by the Physicians for Social Responsibility, a US lobby group that has been awarded the Nobel Peace Prize, says that in addition to polymer chemicals, slurry contains heavy metals such as arsenic and mercury that are common in mined rock. Mine operators usually construct dams to impound the slurry in ponds, or inject it back into closed mines, but both disposal strategies may leach chemicals into groundwater supplies. There's also the risk that slurry dams could burst, as happened in West Virginia in 1972, killing 125 people, while in 2000 a dam burst in Kentucky and entered into local water supplies.[169]

Data for 2011 provided by New Hope to one resident who asked to remain anonymous showed that the Acland mine exceeded its limit for depositional dust on two out of nine occasions. These readings are a crude measure obtained from a glass jar with a funnel on top. For the PM10 readings, only two results were provided over a period of 10 months, and the third reading was missing from the data. The company added that all results for PM10 above the authorised limit of 150 micrograms per cubic metre were reported to the state government, indicating that it had indeed exceeded this limit on an unknown number of occasions. In fact, 150 micrograms is three times the state guideline. No readings were taken for PM2.5. For noise monitoring, the company exceeded its limit in six out of nine readings, but it was able to blame this on other sources: 'insects, wind, fruit bats and dogs barking'. Its compliance status reads 'state of compliance not determinable'.

When Tanya Plant complained to the state environment department about New Hope's poor compliance record in late 2011, she never heard back. Months later, she was told by an official that the department had asked New Hope about the problems and been told that the company was working with the family to resolve them, so the department had taken no further action.

Plant then asked the department why it had not acted on the company's breaches. The same official who had approved New Hope's Environmental Authority to operate defended the department's failure to take action by saying that the system is based on 'complaint driven conditions'. In Orwellian style, the official argued that as no complaints were received – a claim Plant disputes – there were no instances of non-compliance: 'As a complaint was not received near the location that these particular results pertain to there is no applicable non-compliance with these conditions.'[170] If this is how government departments can act when dealing with someone with a PhD from Oxford, then pity the average person in the same situation.

Plant says:

> I feel strongly that [the Department of Environment] has grossly let down my family and others in the district, indeed the whole community, in issuing the Environmental Authority that is currently in operation. It completely abrogates the government or the mine of responsibility for impacts and fails to protect people's health.[171]

Mining companies are spending millions on PR campaigns that emphasise the jobs and revenue they generate, but very little is known or said about the impact on the people living in their path. Governments seem to have made a calculated decision that these people are expendable, and by and large their voices are unheard.

CONCLUSION

MASTER NOT SERVANT OF THE RESOURCES SUPERPOWERS

During the Cold War era, the diplomat Alan Renouf made a confronting assessment of the Australian psyche when he said that fear of our region made us 'The Frightened Country'.[172] Australia has come a long way since then. Successive governments have put in place far-reaching reforms that have made Australia more outward-looking, in turn fostering export-led success that has buoyed economic growth for the past two decades. But Renouf's dismal description could still be applied to the way today's governments are browbeaten by the resources superpowers. These global giants, including BHP Billiton, BG Group, Chevron, Rio Tinto and Xstrata, dominate resource extraction in this country. Individually, their annual income exceeds that of the governments that regulate them, while they are reinforced locally by emerging businesses run by Gina Rinehart, Clive Palmer and Andrew Forrest, among others. State and federal governments are easily intimidated by their threats to withdraw capital and the advertising blitzes that they and their industry associations deploy with alacrity.

The approach taken by Australian governments to the resources sector falls well short of their governance of the rest of the economy. From the mid-1980s onwards, federal governments stared down the powerful banking and manufacturing lobbies and slashed tariffs, deregulated the financial system, floated the dollar, applied a consumption tax to most goods and services, and introduced robust regulation of companies and financial institutions. By contrast, regulation of the resources sector remains in the hands of poorly resourced and often opaque state government departments that lack independence. State governments carry most of the responsibility for regulating these complex projects but, as both regulators and revenue collectors, they remain deeply conflicted. When compared to the institutions that govern the rest of our economy, resource regulation in Australia resembles that found in developing countries. Not that the global giants are complaining.

SUPER CYCLE, SUPER REGULATOR

At a time of soaring investment in mammoth resource projects, federal and state governments and opposition parties have lined up in favour of winding back regulation of resource projects during their approval and operation.

If Australia were struggling to attract investment in such projects, such reform might have merit, but the opposite is manifestly the case. When, in June 2009, the federal environment minister, Peter Garrett, launched the first review of the federal government's main environmental law, the

Environment Protection and Biodiversity Conservation (EPBC) Act 1999, about $120 billion worth of large-scale mineral and energy projects were being built around the country. By the time Garrett's successor, Tony Burke, reported on the review chaired by Dr Allan Hawke, that amount of investment had doubled – and there's another $250 billion of investment at an advanced stage of development.

Tony Burke not only rejected the main thrust of the Hawke review, which called for scrapping the *EPBC Act* and replacing it with a new law, he has also proposed reforms aimed at achieving 'faster environmental assessments' by eliminating so-called red tape to remove duplication and speed up approvals.

Instead of regulatory retreat, the developing super cycle demands stronger and more accountable approval and monitoring processes. Proper protection of the people, land, water and biodiversity affected by big mining projects would be aided by three compelling reforms: boosting the amount of revenue spent on supervising this sector, bolstering the integrity and independence of regulatory authorities, and improving the relationship of federal and state governments.

First, as discussed earlier, the state governments that carry most of the responsibility for regulation poorly fund their relevant institutions. The financial commitment by state and federal governments should be linked to the amount of activity in the sector, as reflected by the revenue they collect. If governments committed to spending an appropriate percentage of their taxes and royalties on regulation and monitoring, then the public might have more confidence that projects were being appropriately supervised. Three to 5 per cent of revenue averaged over a period of three years seems a reasonable amount of insurance to protect against the harmful and long-term effects of a poorly regulated resources sector. Governments could charge back their outlay on a fee-for-service basis and pay competitive salaries to their technical experts, preventing the loss of key talent to the private sector.

Second, the agencies that approve and monitor resource projects should be independent of government ministers, especially state ministers, who mainly have their eyes on the revenue. While some states have independent agencies, they tend to be run by individuals with a strong development bias. The Independent Expert Scientific Committee set up as part of a deal brokered by independent MP Tony Windsor in 2011 should be seen as a starting point for the type of well-resourced, expert institutions we need. With $150 million in funding, its advice will become pivotal for the federal government on CSG and coal projects, while another $50 million is available for advice to state governments. All governments will be required through legislation to take into account the IESC's advice, which will be made public. The states have building blocks to work with, such as the QWC, which produced the CSG groundwater study, but on the whole their regulatory effort is fragmented and ad hoc.

Useful models for future reform are the federal institutions that supervise the money supply, the banking system and corporate activity (the Reserve Bank, the Australian Prudential and Regulation Authority, and the Australian Investments and

Securities Commission). All three institutions are independent of government, engaging high-calibre technical experts and operating under their own independent boards of directors. They don't report to a minister; they report to federal parliament, and they don't receive financial support from the corporate sector. The Reserve Bank, with about 80 professional economists who analyse our complex economy, should be an aspirational model for federal and state ministers if they are interested in harnessing the level of expertise needed.

A strong financial commitment by government would help to make regulation more independent. Corporate contributions, either through universities or independent studies, are designed to achieve favourable outcomes. This was demonstrated by the recent Namoi Water Study, which was partly funded by the federal government in response to concerns among farmers in the Liverpool Plains about proposed coal mining and CSG projects. While the government kicked in $1.5 million, industry funding dominated the initiative with a $3 million contribution. This contribution seemed to achieve the desired results. The company chosen to do the study was a subsidiary of the US-based oil services firm Schlumberger, which also works on CSG projects in Australia. Farm groups in the region say the scope of the study did not reflect the terms of reference, and that it left out water quality and was designed to minimise the impact of the proposed projects.[173]

Third, a federal–state institution seems the best way of eliminating duplication and bolstering the regulatory wherewithal needed, especially when it comes to dealing with impacts that flow across state borders. In the case of corporate regulation, state institutions were found wanting after the corporate collapses of the late 1980s. The states referred their powers to the federal government, which introduced the Corporations Law and created the Australian Securities and Investments Commission (ASIC). Unlike in the case of corporate collapse, the results of poor regulation of resource projects could be irreversible, so governments need to act before things go wrong. A useful model for future reform is NOPSEMA, the new national offshore oil and gas regulator, which also operates as an independent statutory authority.

At present, Environmental Impact Statements are patchy and generally biased towards development: that is why so many projects get approved. Independent EISs managed by an independent regulator are the only way to ensure that the public gets a full and frank account of the risks. Instead of the companies choosing their pet consultants to produce statements, the regulator should manage the process independently and charge the cost back to the company. The consultants who produce these reports should be accredited by the independent regulator. Those caught producing dodgy data will find themselves struck off the list, in the same way that ASIC strikes out directors and financial advisers.

REVIEW TAXATION

At the heart of the resources stampede are the relatively low rates of taxes and royalties levied on companies that extract our non-renewable natural resources. Australian governments don't administer taxes that take into account the finite nature of these resources, nor do they charge foreign and local companies for the privilege of operating in a nation that follows the rule of law and has democratic institutions.

Instead, the mining companies treat us like fools when they run full-page newspaper ads that claim their tax contribution has risen by 500 per cent. Such ads rely on the sheer dollar amounts paid, which is a meaningless measure; the only true measure is taxes and royalties as a percentage of profits, and this indicates fairly low rates. According to Treasury modelling, only for very profitable mines do Australian taxes on iron ore exceed those of Brazil, while taxes on coal are lower than those of Indonesia for all but the exceptionally profitable mines.

Australia can and should charge resources companies a premium for profiting from the first world governance which presides over extraction of our finite resource wealth. While the companies claim they are subject to tax rates in the 40 per cent range, Reserve Bank data indicates that taxes and royalties together amount to about 27 per cent of net revenue.[174] Compare that to the savvy Norwegians, who charge local and foreign companies a nominal rate of 70 per cent for exploiting their oil, and don't have trouble attracting foreign investment. While I'm not advocating this rate, we should be looking at profits-based taxation that collects a fair share when the resources companies hit paydirt, without discouraging investment. A less generous and less exploitable resource rent tax than the one we now have would achieve this.

Evidently, a calm, considered and comprehensive review of the taxation of resource projects is sorely needed. This should include the botched mineral resource rent tax (MRRT), which only applies to coal and iron ore and is a much watered-down version of the resource super-profits tax. The loopholes in the MRRT are so exploitable that tax planners could drive a 300-tonne mining truck through them. Companies can choose the most favourable depreciation regime, in the manner of the US-based iron ore miner Cliffs Natural Resources, which has used this device to create a $300 million tax shield. The global giants, with multiple projects in Australia, will be able to offset their MRRT liability against the development costs of their ever-expanding mines. At the same time, the MRRT puts an onerous burden on the local emerging miners that do not have the size or scope to minimise their obligations. As a result, Treasury modelling indicates that a local iron ore miner will pay a tax rate of close to 50 per cent.

The tax review should also include state governments, which have clung to royalties because the federal–state system denies them broad-based taxes. These production-based royalties have been inherited from Britain in the Middle Ages. All of Australia's six states rely on such royalties, some of which are levied at cents-per-tonne rates, which means they take no account of the value of the resource. Others are levied as a share of production, which means they fail to capture a share of super profits. Mining companies have been successful in stamping out royalties designed to collect a share of profits, such as the original agreement for Olympic Dam negotiated in the early 1980s. When the South Australian government negotiated in 2011 a new 50-year agreement with BHP for the $30 billion open-cut expansion, its hapless advisers and ministers didn't insist on a profits-based royalty, thereby denying their citizens a direct dividend from what's likely to become a very profitable mine.

Royalties are an inefficient way of raising money. They penalise start-up operators – mostly local companies – because the impost must be paid as soon as production starts, thereby giving a distinct advantage to the big multinationals that can easily fund them.

Royalties also encourage unethical behaviour because they bias our political process towards development. In fact, they now seem to operate like inducements paid by mining companies to get their projects over the line. Rarely do you hear companies complaining about having to pay royalties – there's been no push by the industry to have them abolished and replaced with profits-based taxes. Why? Because royalties bring governments inside the tent with the miners, and this is wrong. Governments are elected to govern for all of us, but with royalties their judgment is clouded. These imposts should be gradually phased out and replaced with profits-based taxes that will return a greater share of revenue to Australian citizens over the life of this mining boom, and to future citizens who are going to have to live with the consequences of our insatiable need for resource projects.

TREASURE FOOD-PRODUCING LAND

The Hunter Valley experience shows how our current governance model tips the balance between resource development and farming/conservation towards extreme development.

In a world with a growing population, it seems contrary to Australia's national interest to sacrifice food-producing land for short-term gain. The simple measure of land productivity, combined with access to groundwater, could be used to decide on the regions that must be fully protected from mining and CSG wells for present and future generations of food producers.

Coexistence, as attempted in the Hunter Valley, doesn't work, simply because mining companies will always start with a minimalist mine and a view to making it a mega-mine, should commodity prices justify the investment. It should be the job of an independent regulator to devise a system to protect our food bowls, based on extensive consultation, followed by strict monitoring and enforcement. Farmers operating in these designated agricultural areas should have the right to say no to mining and CSG companies coming onto their land, including through exploration.

Australia needs to take a strategic approach to food production. In 50 years, the growing populations of Asia might find renewable alternatives to our coal and gas, but they will almost certainly be importing greater amounts of our high-quality grain and protein.

NO SECRETS

Resources companies rely on strict confidentiality agreements with landholders, including Indigenous communities, to increase their negotiating power. They also rely on secrecy when reporting their compliance with the conditions prescribed in government approvals. Yet secrets only serve the interests of the powerful, and state governments should legislate to make these agreements publicly available in an online register. Real-time public reporting on compliance should also be mandatory and available on websites. The federal Opposition resources spokesman, Ian Macfarlane, who is both a farmer and former resources minister, has pledged to introduce full disclosure of all relevant data by resources companies. This policy is worthy of bipartisan support, and should be extended to every state government.[175]

At the heart of Indigenous disadvantage in Australia are the highly secretive negotiations forced on communities by resources companies. Again, secrecy does not serve the interests of these people, who often negotiate with little understanding of what a fair deal looks like. As discussed in 'The Tail That Wags the Dog', in 2002 Tony Blair launched the Extractive Industries Transparency Initiative. The failure of the federal government to introduce the EITI regime simply confirms that the global giants have more sway here than they do in other resource-rich countries.

MAKE MINES NORMAL WORKPLACES

Working conditions in mines are now pushing the boundaries of acceptability. In the space of a few years, more than a century of progress in the workplace has been overturned by shifts of 12 hours or more that extend for two weeks or more at a stretch. The FIFO/DIDO regime is putting individuals, families and communities under extreme pressure and government should act to stem the expansion of this model.

Resources companies should be made partly responsible for the health and safety issues that arise from commuting to and from their sites. There can be no denying that drive-in, drive-out arrangements are leading to greater numbers of road accidents, including a rising number of fatalities, as a result of the dangerous combination of 12 hour–plus shifts and long distances.

Companies should also take some responsibility for the rising number of relationship breakdowns caused by these arrangements, and the impact on children left without a parent for extended periods. Governments should get involved by removing disincentives to locate workers near mines, such as the fringe benefits tax, which creates a tax liability when companies provide accommodation for workers. The cost of providing a donga in a work camp is fully tax-deductible, but when the company provides a normal home to a worker, this is subject to FBT. Changes to the fringe benefits tax could be paid for by putting a surcharge on FIFO flights. Governments and companies should encourage the rapid roll-out of more flexible and inexpensive housing in boom towns. More affordable housing would help to facilitate a much-needed shift back to locating workers and their families in mining regions. If Australia were at war, the work arrangements that now apply in the resources industry would be warranted, but this is clearly not the case. At the heart of a more sensible approach to workplace reform should be a return to a civilised eight-hour day.

PROTECT AND COMPENSATE

The toll inflicted on communities and landholders living in the shadow of mines and CSG fields already seems intolerable, yet there will be many more people forced to live in the fallout zone of ever-expanding projects as this boom rolls on. One of the solutions must involve independent monitoring of the dust, noise and gases released by these projects, and strong enforcement to curb production when these limits are exceeded.

Even with these measures in place, people living near mines and major gas facilities should be given the option of fair compensation should they wish to leave. Proximity is an open question, but a distance of five kilometres seems reasonable, though the right to compensation could be dependent on transparent benchmarks such as demonstrable health effects. Such compensation should be based on the land value before the mine obtained its exploration licence, and the valuations should average the results of independent valuers selected by both the mine and the resident.

What happened to the people of Acland on Queensland's Darling Downs should never be repeated in this country. In this case, a mining company went about buying up farms and homes before it had received approval for its expansion. The Stage 3 approval for New Hope's open-cut mine was then rejected by the incoming LNP state government, but Acland is now a virtual ghost town with one permanent resident, while the carcasses of removed homes lie dumped in a paddock. State governments should legislate against property acquisitions by mining and energy companies for developments yet to be approved.

GAS IN THE NATIONAL INTEREST

It seems a very poor outcome indeed: the development of Australia's CSG reserves threatens not only to damage our prime farmland and groundwater resources, but also to double or triple domestic prices because export sales are 'hoovering up' all the available gas. And all of these new exports will drive up the Australian dollar as well, crunching the non-mining economy. Yet the federal government seems hell-bent on even greater export 'success' – it wants to make Australia the biggest gas exporter in the world, even though our reserves are ranked twelfth in size.[176]

Developing our gas resources at breakneck speed, with so many inherent risks, does not seem to be in the national interest. Evidently, Australia needs a policy to develop our resources while protecting domestic supply. The United States has a policy of reserving its gas for the domestic market before allowing exports. The federal Opposition has expressed support for a reservation policy, which would only be applied prospectively and is worthy of bipartisan support.[177] When applied to large-scale projects that need export markets to become economical, the government could simply ask for a set percentage to be allocated to the domestic market.

The CSG rush in Australia has come about largely because a state government saw an easy opportunity for a quick revenue grab. A more thorough and long-term approach would have seen governments encourage gas companies to develop Australia's more remote gas resources, which would not threaten our food bowl. This is what our governments did in the 1960s and 1970s when we built the Moomba pipelines from the remote northeast corner of South Australia to Sydney and Adelaide.

Like the United States, Australia has potentially vast amounts of shale gas, but gas companies haven't looked at these resources because they've been allowed to grab the resources close to the coast. Australia's shale gas resources, estimated by Geoscience Australia at 435,000 petajoules, are more than double our CSG reserves and four times those of the northwest shelf.[178] These reserves are located in remote Queensland, South Australia, Western Australia and the Northern Territory. These areas should be Australia's next frontier.

ACKNOWLEDGEMENTS

First and foremost I must thank the entire team at Black Inc. who conceived of this book as a sequel to *Too Much Luck*, rather than the mere update I had in mind. I must especially thank Chris Feik who provided steady guidance, and Kate Goldsworthy who worked on the manuscript and managed to put things where I had originally intended them to be, and Elisabeth Young and Phoebe Wynne.

Aside from those who made this book possible by agreeing to be interviewed and quoted, I'd like to thank the following people who have supported my research or indeed inspired this work: Heather Brown and David Pascoe, Nikki and Glen Laws, Rob McCreath, Lee McNicholl, Robert Murray, Glen and Gary Slee, Velocity Village Karratha, Trent Deverell, Peter Colley and Séan Kerins. Nick Cater, editor of the *Weekend Australian*, also inspired this work during a conversation we had following his return from a visit to the Darling Downs in late 2011. Chris Mitchell and Clive Mathieson, editor-in-chief and editor of the *Australian* respectively, have given me the opportunity to write about the issues which I have treated in greater detail in this book. Staff at DERM/DEHP answered my numerous questions in a professional way. And a very big thank you to the numerous people who contributed to this book but can't be named for professional reasons.

The Centre for Aboriginal Economic Policy Research at the Australian National University assisted with fieldwork in the Northern Territory in 2011, as did staff at the Northern Australia Research Unit, especially Nicole Everett, and Arlo. ANU maintains a wonderful facility for researchers wanting to work in this important part of Australia; may it always do so. Jon Altman, Bob Gregory and Colin Filer have encouraged me to think deeply and critically about the effect the resources boom is having on Australia.

I'd also like to thank friends, family and colleagues who have encouraged me along the way, especially Paul Daley, Eleanor Hall, Lenore Taylor, Tom Dusevic, Steve Lewis, Matthew Kelly, Margaret Pancino, Teresa and Barry Price, Anne Blackall and my agent, Lyn Tranter. I am also indebted to those who provided encouraging feedback on *Too Much Luck*, which made me realise that this new book was necessary. Any feedback on this book can be directed to orestruck@gmail.com.

Finally, a big thank you to Kate and Patrick Lio for putting up with my long spells locked away and for taking part in some of the fieldwork.

ENDNOTES

1. The Australian Bureau of Statistics' national accounts data for March
    2012 estimates the gross value-added output of the mining and oil sector
    at 7.5 per cent of GDP, which is often quoted as the industry's size. This
    figure excludes the inputs to the resources sector from other industries.
    Mining revenue includes these inputs, and shows the direct and indirect
    size of the mining industry. These figures have been taken from Connolly,
    E. and Orsmond, D. *Research Discussion Paper, 2011–08: The Mining
    Industry: From bust to boom*, Sydney: Reserve Bank of Australia,
    December 2011.
2. Author's estimate based on statements made by iron ore companies.
    Explosives maker Orica announced on 22 May 2012 that it would invest
    $814 million in a new ammonium nitrate plant in Western Australia. The
    decision was underpinned by Orica's estimate of 1.1 billion tonnes of iron
    ore production in Western Australia by 2020, given that iron ore mining
    requires a lot of explosives.
3. I am indebted to Nick Cater who first made this observation to me.
4. Department of Employment, Economic Development and Innovation.
    Guide to Queensland's New Land Access Laws, Brisbane: The State of
    Queensland, November 2010, pp. 4–5.
5. Interview with author, 20 December 2011.
6. Email response to author, 29 June 2012.
7. *Ibid.*
8. Bureau of Resource and Energy Economics. *Mining Industry Major
    Projects*, Canberra: Australian Government, April 2012.
9. Interview with author, 27 May 2012.
10. ABC News Online. 'Palmer blasts "poisonous" coal seam gas industry',
    29 August 2011.
11. Waratah Coal. 'Cumulative Impact Assessment', *China First Environmental
    Impact Statement*, Volume 1, Chapter 5, 2011, p. 66.
12. Waratah Coal. 'Project Overview', *China First Environmental Impact
    Statement*, Volume 1, Chapter 1, pp. 2–6.

13. Franks, D., Brereton, D. and Moran, C.J. 'Surrounded by Change: Collective strategies for managing the cumulative impacts of multiple mines', paper presented to SDIMI conference, Gold Coast, 6–8 July 2009.

14. *Ibid.*, p. 60.

15. O'Faircheallaigh, C., address to Catalyst Conference, Sydney, 14 November 2011.

16. ABC News Online. 'Palmer blasts "poisonous" coal seam gas industry'.

17. Email response, Department of Employment, Economic Development and Innovation, 12 April 2011.

18. Email response, NSW Trade and Investment, 20 February 2012.

19. Edwards, G. 'Is There a Drop to Drink? An issues paper on the management of water co-produced with coal seam gas', Queensland: Department of Mines and Energy, 2006, pp. 53–67.

20. *Ibid.*, pp. 53–4.

21. Cleary, P. 'Claims of power price signalling', *The Australian*, 24 November 2012.

22. CSIRO. 'What Is Coal Seam Gas?', 29 November 2011, p. 1.

23. CSIRO. 'Coal-Seam Gas – Produced water and site management', Melbourne, November 2011, p. 1.

24. National Water Commission. 'The Coal Seam Gas and Water Challenge: National Water Commission position', Canberra, 3 December 2010, nwc.org.au.

25. Moore, A. 'Every CSG aquifer must be tested: Burke', *Lateline*, ABC Television, 30 August 2011.

26. Moran, C. and Vink, S. 'Assessment of Impacts of the Proposed Coal Seam Gas Operations on Surface and Groundwater Systems in the Murray-Darling Basin', Brisbane: Sustainable Minerals Institute, 29 November 2010, pp. 4–5.

27. Garrett, P. 'Attachment B, letter to Penny Wong', 12 July 2010, included as an appendix in a submission from the Department of Sustainability and Environment to Tony Burke, 27 September 2010. Obtained by author under FOI law.

28. Hillier, J.R. 'Groundwater Connections between the Walloon Coal Measures and the Alluvium of the Condamine River: A report for the Central Downs Irrigators Ltd', August 2010, p. 21.

29. Queensland Water Commission. 'Draft Underground Water Impact Report for the Surat Cumulative Management Area: Consulatation draft', Brisbane, May 2012, pp. 54–7.

30. *Ibid.*, pp. 1 and 28.

31. Osborn, S.G. et al. 'Methane contamination of drinking water accompanying gas-well drilling and hydraulic fracturing', *Proceedings of the National*

Academy of Sciences of the United States of America, 17 May 2011, Volume 108, Number 20, pp. 8172–6.

32. Email statement, 21 March 2012.

33. CSIRO. 'CSG developments – Predicting impacts', Melbourne, April 2011, p. 2; CSIRO. 'Coal-Seam Gas – Produced water and site management', p. 2.

34. *Ibid.*, p. 4.

35. Carlisle, W. 'Queensland reveals Condamine water quality report', *Coal Seam Gas: By the Numbers*, ABC Radio Multiplatform and Content Development, ABC News Online, 2012.

36. Statement to author from the office of the Hon. Jeff Seeney, 13 June 2012.

37. AAP. 'CSG breaches will be punished, says Cotter', 20 April 2012.

38. Klan, A. 'Concern looms over chemicals left behind', *The Australian*, 2 December 2011.

39. US EPA. 'EPA Releases Draft Findings of Pavillion, Wyoming Ground Water Investigation for Public Comment and Independent Scientific Review', 8 December 2011, yosemite.epa.gov.

40. CSIRO. 'Hydraulic fracturing (fraccing)', December 2011, p. 1.

41. *Ibid.*

42. National Toxics Network. 'Hydraulic Fracturing in Coal Seam Gas Mining: The Risks to Our Health, Communities, Environment and Climate', Bangalow, 2011, p. 10.

43. *The Economist.* 'We will frack you: The rise of shale gas continues', 19 November 2009.

44. Klan, A. 'Resources veteran takes aim at CSG "cowboys"', *The Australian*, 5 December 2011.

45. Department of Sustainability and Environment. 'Proposed Decisions – Australia Pacific LNG Project', 2 February 2011, p. 2. Obtained by author under FOI law.

46. Department of Employment, Economic Development and Innovation. *Land Access Code*, Brisbane: The State of Queensland, November 2010, p. 9.

47. Howarth, R.W., Santoro, R., and Ingraffea, A. 'Methane and the greenhouse-gas footprint of natural gas from shale formations – A letter', *Climatic Change*, Volume 106, Number 4, 2011, pp. 679–90.

48. Statement to author by Origin spokesman, 11 May 2012.

49. Statement to author by QGC spokesman, 4 April 2012.

50. Interview with author, 1 February 2012.

51. Interview with author, 26 March 2012.

52. Interview with author, 17 March 2012.

53. *Ibid.*

54. *Ibid.*

55. Oatley, S. 'A clear abuse of the law', *The Australian*, 11 June 1996.

56. 'Editorial', *The Sydney Morning Herald*, 11 June 1996.
57. *Ibid.*
58. Oatley, 'A clear abuse of the law'.
59. NSW Department of Trade and Investment, Regional Infrastructure and Services. *NSW Coal Industry Profile 2010: Incorporating Coal Services Pty Ltd*, Sydney: NSW Government, 2011, p. 230.
60. Kirkwood, I. 'Vineyard swallowed by BHP Billiton mine', *The Newcastle Herald*, 16 October 2009.
61. Interview with author, 19 March 2012.
62. Souris, G. Debates, Legislative Assembly, Parliament of New South Wales, Number 21, 12 June 1996.
63. Beder, S. 'Market forces and the surrender of professional judgment', *Engineers Australia*, August 1998, p. 62.
64. Interview with author, 17 March 2012.
65. Letter from Mary Tolson to Xstrata Coal, 1 November 2011.
66. Biographical details from *Encyclopedia of Australian Science*, accessed via http://trove.nla.gov.au/people.
67. Shepherd, N. et al. 'Warkworth Extension Project (09_0202)', Sydney: NSW Department of Planning, 9 February 2012, p. 8.
68. Guilliatt, R. 'What they're saying is that we aren't worth anything', *The Weekend Australian Magazine*, 26–27 May 2012.
69. Letter from Tolson to Xstrata, quoting material presented by the company at a 'visioning' workshop, 1 November 2011.
70. Telephone conversation with author, 22 March 2012.
71. Letter from Xstrata to Mary Tolson, 25 November 2011.
72. Letter from Tolson to Xstrata, 1 November 2011.
73. *Buzzacott v Minister & Ors* (No. 2) [2012] FCA 403.
74. Council of the Australian Governments Meeting. *Communiqué*, Canberra, 13 April 2012, p. 2.
75. NSW Environment Protection Authority. 'EPA issues penalty notice to Whitehaven Coal', Media Release, 6 March 2012.
76. NSW Environment Protection Authority. 'Moolarben coal fined $105,000 for pollution of waterways', Media Release, 30 March 2012.
77. Behrmann, E. 'Mining truck tires pricier than porsches, Miami condominiums', Bloomberg, 30 June 2011, www.bloomberg.com.
78. 'About' page, Peabody Energy Australia, www.peabodyenergy.com.au.
79. ABC Online. 'Australian mines top global investment list', 27 March 2012.
80. Email obtained by author.
81. Scott, S., AAP. 'LNP's approach to approving Alpha Coal project labelled "shambolic"', *The Courier-Mail*, 5 June 2012.
82. Email to author, 27 April 2012.

83. Department of Mines and Petroleum. *Annual Report, 2010–11*, Perth: Government of Western Australia, pp. 69–70.

84. Planning Assessment Commission. *Annual Report, 2010/2011*, Sydney: NSW Government, p. 43.

85. Email response to author.

86. Email response to author, 16 February 2012.

87. Email statement to author, 5 April 2012.

88. Bonyhady, T. 'Introduction', in Bonyhady, T. and Macintosh, A. (eds). *Mills, Mines and other Controversies: The environmental assessment of major projects*, Annandale: The Federation Press, 2010.

89. Elliott, M. and Thomas, I. *Environmental Impact Assessment in Australia: Theory and practice*, Annandale: The Federation Press, 5th Edition, 2009, p. 127.

90. Email response from the Department of Environment, 16 March 2012.

91. *Ibid.*

92. *Ibid.*

93. Howey, K. 'Northern Territory and the McArthur River Mine', in *Mills, Mines and other Controversies*, p. 69.

94. McLaughlin, M. 'Garrett flies in to NT to sort out mining dilemma', ABC Television, 15 January 2009.

95. Interview with author, August 2011.

96. McLaughlin, 'Garrett flies in to NT to sort out mining dilemma'.

97. Email response from departmental spokeswoman, 16 March 2012.

98. Interview with author, 8 March 2012.

99. Interview with author, 22 February 2012.

100. Environmental Defender's Office (Ltd) NSW. *Ticking the Box: Flaws in the Environmental Assessment of Coal Seam Gas Exploration Activities*, Sydney, November 2011, p. 20.

101. Environmental Defender's Office (Qld) Inc and Environmental Defender's Office of NQ. 'Mining and Coal Seam Gas: Access to information, fair process and the Land Court', submission to Premier Anna Bligh and ministers, 15 June 2011, p. 2.

102. Elliott and Thomas. *Environmental Impact Assessment in Australia*, p. 264.

103. Email from Alex Bruce to Richard Sheldrake, 25 May 2011.

104. Interview with author, 21 March 2012.

105. Interview with author, 3 April 2012.

106. Reilly, T. 'The fortune and fury of a young tycoon', *The Sunday Age*, 10 October 2010.

107. Interview with author, 21 March 2012.

108. Cullen, S. '"Green veto" holding back Qld coal plan: Abbott', ABC News Online, 13 June 2012.

109. Crowe, D. 'Keep taxes stable or lose boom: Martin Ferguson', *The Australian*, 15 May 2012.

110. Cleary, P. 'Clan fury as MP clears hurdle for friend Andrew Forrest', *The Australian*, 21 December 2011.

111. Bita, N. 'Campbell Newman slams farm gate shut on miners', *The Australian*, 29 March 2012.

112. *Ibid.*

113. Ingraffea, A. 'Does the natural gas industry need a new messenger?', *Fractured Future: A series of special op-eds about the shale gas industry*, CBC News, 29 November 2011.

114. *Ibid.*

115. Email to author from Andrew Potter, 30 May 2012.

116. Email response from Jan King, 13 April 2012.

117. Interview with author, 3 April 2012.

118. Interview with author, 19 October 2011.

119. Interview with author, 22 March 2011.

120. Langton, M. 'The resource curse', *GriffithREVIEW: Still the Lucky Country?*, Edition 28, 2010, p. 48.

121. *Ibid.*, p. 56.

122. Gray, M. and Langton, M. 'Resource companies deliver the goods', *Australian Financial Review*, 28 January 2012, p. 45.

123. Altman, J. and Biddle, N. 'Rudd overpromised on Indigenous unemployment', *Crikey*, 4 June 2010. Altman is one of the author's PhD supervisors.

124. Cousins, D. and Nieuwenhuysen, J. *Aboriginals and the Mining Industry: Case Studies of Australian Experience*, Sydney: George Allen & Unwin, 1984, pp. 167–8.

125. Confidential author interview with Indigenous employee.

126. Author interview with managers and employees of Yolngu Businesses Enterprises, August 2011.

127. Centre for Aboriginal Economic Policy Research. *People on Country, Healthy Landscapes and Indigenous Economic Futures: 2011 Annual Report*, Canberra: Australian National University, College of Arts and Social Sciences, 2012, p. 17.

128. Jordan, K. 'Twiggy v Gillard on Aboriginal jobs: Who's really delivering?', *Crikey*, 28 May 2012.

129. Email to author, 4 June 2012.

130. Interview with author, 15 February 2012.

131. Savas, K. 'Affidavit of Kyriakos Savos Sworn 9 March 2012', Federal Court of Australia, Western Australia District Registry, 2012, p. 5.

132. Interview with author, 15 May 2012.

133. Obama, B. 'Opening Remarks by President Obama on Open Government

Partnership', Washington DC: The White House, Office of the Press Secretary, 20 September 2011.

134. Barnett, C. 'Release of comprehensive development strategy for Burrup Peninsula', Media Release, 16 October 1996.

135. Collier, P. '5392, Hon. Robin Chapple to the Minister for Indigenous Affairs, 2 May 2012', Questions, Legislative Council, Parliament of Western Australia, Number 202, 29 March 2012, p. 2.

136. Email response from Laura Lunt, Woodside spokeswoman, 18 April 2012.

137. Interview with author, 4 March 2012.

138. Interview with author, 12 March 2012.

139. Accident statistics provided to author by Queensland Department of Transport and Main Roads, 3 April 2012.

140. Interview with author, 16 December 2011.

141. Hennessey, A. 'Findings into the deaths of Malcolm MacKenzie, Graham Brown, and Robert Wilson', *Office of the State Coroner: Finding of Inquest*, Coroner's Court, Rockhampton, 23 February 2011, pp. 1–2.

142. Laurie, V. 'Flying into Trouble', *The Weekend Australian*, 7–8 April 2012.

143. Interview with author, 3 March 2012.

144. Deloitte Access Economics. *2020 Growth Outlook Study*, Queensland Resources Council, p. 34.

145. Laurie, 'Flying into Trouble'.

146. WA Department of Racing, Gaming and Liquor, 'Find a Licence' tab on website.

147. Interview with author, 16 October 2011.

148. Carrington, K. et al. 'Inquiry into the use of "fly-in, fly-out" (FIFO)/ Drive-in, Drive Out (DIDO) workforce practices in regional Australia', Submission to the APH House of Representatives Inquiry, Number 95, Canberra: Standing Committee on Regional Affairs, 9 October 2011, p. 15.

149. *Ibid.*, pp. 22–23.

150. Carrington, K., Hogg, R. and McIntosh, A. (2010), 'The resource boom's underbelly: Criminological impacts of mining development', *Australian & New Zealand Journal of Criminology*, December 2011, Volume 44, Number 3, p. 342.

151. Confidential response to survey by Kerry Carrington.

152. Carrington et al. 'Inquiry into the use of "fly-in, fly-out" (FIFO)/Drive-in, Drive Out (DIDO) workforce practices in regional Australia', p. 9.

153. Edited announcement. 'State calls on mining companies to clean up act', *Western Australian Business News*, 13 November 2007.

154. Interview with author, 8 November 2011.

155. Cleary, P. 'Miners strain regional medical facilities', *The Australian*, 17 October 2011.

156. Interview with author, 10 November 2011.

157. Interview with author, 10 November 2011.

158. Diary shown to author, 10 November 2011.

159. Interview with author, 16 January 2012.

160. Queensland Department of Transport, statement to author, 7 December 2012.

161. RailCorp, email statement to author, 5 December 2012.

162. Neale, R. 'Letter to the Editor', *The Australian*, 1 February 2012.

163. SA Health. 'Technical Paper 2012/1: Analysis of blood lead levels for the first quarter of 2012 (31 March 2012)', Adelaide: Government of South Australia, p. 4.

164. McKenna, M. 'The toll lead takes on Mt Isa's infants', *The Australian*, 17 September 2010.

165. Email response, 5 April 2012.

166. Population Health Division, NSW Department of Health. *Respiratory and cardiovascular diseases and cancer among residents in the Hunter New England Area Health Service*, North Ryde: NSW Department of Health, 2010.

167. Au, T., address to Lock the Gate Alliance at the Forum, Sydney, 9 March 2011.

168. Hendryx, M. and Ahern, M. 'Relations Between Health Indicators and Residential Proximity to Coal Mining in West Virginia', *American Journal of Public Health*, Volume 98, Number 4, April 2008, pp. 669–71.

169. Lockwood, A. et al. *Coal's Assault on Human Health*, Washington DC: Physicians for Social Responsibility, November 2009, p. 7.

170. Email to Tanya Plant from DERM official, 8 March 2012.

171. Email to author, 16 March 2012.

172. Renouf, A. *The Frightened Country*, Melbourne: Macmillan, 1979.

173. Namoi Community Network. 'Community goes public on Namoi Catchment Water Study concerns', Media Release, 6 March 2012; Maules Creek Community Council. 'Namoi Water Study falls short of the mark', Media Release, 2 April 2012.

174. Connolly and Orsmond. *Research Discussion Paper, 2011–08*, p. 31. The RBA defines net revenue as the residual after paying for the labour and other intermediate inputs used in mining operations; the balance of the revenue (termed the gross operating surplus) is divided between royalty and tax payments, immediate dividend distributions and retained earnings (after deducting interest and depreciation).

175. Cleary, P. 'Libs demand more from developers', *The Australian*, 14 May 2012.

176. Central Intelligence Agency. 'Natural Gas – Proved reserves', *The World Factbook*, www.cia.gov.

177. Ian Macfarlane, interview with author, 1 June 2012.
178. Bradshaw, M. et al. *Australian Gas Resource Assessment 2012*, Canberra: Department of Resources, Energy and Tourism, Geoscience Australia and Bureau of Resource and Energy Economics, p. 3.

NOTES ON MAPS

Map 1, p. 11. Operating mine locations:
Geoscience Australia data derived from www.australianminesatlas.gov.au for all commodities and all operating mines. Dot sizes derived from website.

Map 2, p. 20. Coal-seam gas bore locations:
Queensland: Based on or contains data provided by the State of Queensland (Department of Employment, Economic Development and Innovation) 2011 which gives no warranty in relation to the data (including accuracy, reliability, completeness or suitability) and accepts no liability (including without limitation, liability in negligence) for any loss, damage or costs (including consequential damage) relating to any use of the data.

New South Wales: © Copyright 11 in this information (data) is vested in the Crown. The State of New South Wales retains ownership of its own intellectual property rights. Short quotations from the text of this publication and copies of maps, figures, tables, etc (excluding any subject to pre-existing copyright) may be used in scientific articles, exploration reports and similar works provided that the source is acknowledged and subject to the proviso that any excerpt used, especially in a company prospectus, Stock Exchange report or similar, must be strictly fair and balanced. Other than for the purposes of research or study the whole work must not be reproduced without the permission in writing of the New South Wales Department of Primary Industries.

Maps 1 and 6, p. 11 and p. 49. Mine and deposit locations:
© Commonwealth of Australia 2000. This work is copyright. Apart from any fair dealings for the purposes of study, research, criticism or review, as permitted under the Copyright Act, no part may be reproduced by any process without written permission. Copyright is the responsibility of the Chief Executive Officer, Australian Geological Survey Organisation. Inquiries should be directed to the Chief Executive Officer, Australian Geological Survey Organisation, GPO Box 378, Canberra City, ACT, 2601 AGSO has tried to make the information in this product as accurate as possible. However, it does not guarantee that the information is totally accurate or complete. Therefore you should not rely solely on this information when making a commercial decision.